My Book

This book belongs to

Cover Design by :
Gowri Vemuri

First Edition :
January , 2020

SecondEdition :

April , 2021

Author :
Gowri Vemuri

Edited by :
Raksha Pothapragada
Ritvik Pothapragada
Questions: mathknots.help@gmail.com

NOTE : VDOE is neither affiliated nor sponsors or endorses this product.

Dedication

This book is dedicated to:

My Mom, who is my best critic, guide and supporter.

To what I am today, and what I am going to become tomorrow,

is all because of your blessings, unconditional affection and support.

This book is dedicated to the

strongest women of my life ,

my dearest mom

and

to all those moms in this universe.

G.V.

Visit www.a4ace.com

Also available more time based practice tests on subscription

The Virginia Board of Education and Virginia Department of Education (VDOE) have developed the Virginia Assessment Program (VAP) to measure and evaluate students' academic progress in the Standards of Learning (SOLs). The SOLs indicate Virginia's expectations for what students should know and be able to do in the subject areas of reading, writing, mathematics, science, and history/social science.

Students in grades 3-12 to take the Standards of Learning (SOL) assessments each year. Some of the tests are required by all students each year, and others are required only at specific grade levels. Additionally, with the removal of some SOL tests in recent years, the VDOE assigned the responsibility of the creation and administration of alternate, performance - based assessments on local divisions. Student scores from these tests determine a school's and the division's state accreditation and measures progress toward meeting federal targets.

Virginia Standards of Learning (SOL) tests are generally given online unless a student has an identified and documented need to be assessed using paper,pencil format. The test question format is typically multiple choice, and each test contains some technology enhanced items.

GRADE 3-8 STANDARDS OF LEARNING(SOL) TESTS

GRADE 3	GRADE 4	GRADE 5	GRADE 6	GRADE 7	GRADE 8
	VIRGINIA STUDIES				WRITING
MATH	MATH	MATH	MATH	MATH	MATH
READING	READING	READING	READING	READING	READING
		SCIENCE			SCIENCE

NOTE : VDOE is neither affiliated nor sponsors or endorses this product.

END OF COURSE STANDARDS OF LEARNING (SOL) TESTS

GRADE 9	GRADE 10	GRADE 11
ALGEBRA I	GEOMETRY	ALGEBRA II
EARTH SCIENCE	BIOLOGY	CHEMISTRY
WORLD HISTORY I	WORLD HISTORY II	VIRGINIA & US HSTORY
		WORLD GEOGRAPHY
		ENGLISH : READING
		ENGLISH : WRITING

Any Student taking one of the courses listed here is expected to take the corresponding end-of-course SOL test. The grade levels depicted here represent grade level at which students typically participate in these courses.

NOTE : VDOE is neither affiliated nor sponsors or endorses this product.

SOL Test Scoring and Performance Reports:

Standards of Learning assessments in English reading, mathematics, science and history/social science are made up of 35-50 items or questions that measure content knowledge, scientific and mathematical processes, reasoning and critical thinking skills. English writing skills are measured with a two-part assessment that includes multiple-choice items and an essay.

Student performance is graded on a scale of 0-600 with 400 representing the minimum level of acceptable proficiency and 500 representing advanced proficiency. On English reading and mathematics tests, the Board of Education has defined three levels of student achievement: basic, proficient, and advanced, with basic describing progress towards proficiency.

Performance Achievement Levels:

- The achievement levels for grades 3-8 reading and mathematics tests are: *Pass/Advanced, Pass/Proficient, Fail/Basic,* and *Fail/Below Basic.*

- The achievement levels for science tests, history tests, and End-of-Course (EOC) tests* are: *Pass/Advanced, Pass/Proficient*, and *Fail/Does Not Meet.*

- The EOC Writing (2010 SOL) test, EOC Reading (2010 SOL) test, and EOC Algebra II (2009 SOL) test have an achievement level of *Advanced/College Path* in place of the *Pass/Advanced* achievement level.

NOTE : VDOE is neither affiliated nor sponsors or endorses this product.

INDEX

FORMULA SHEET

1. Area of a triangle

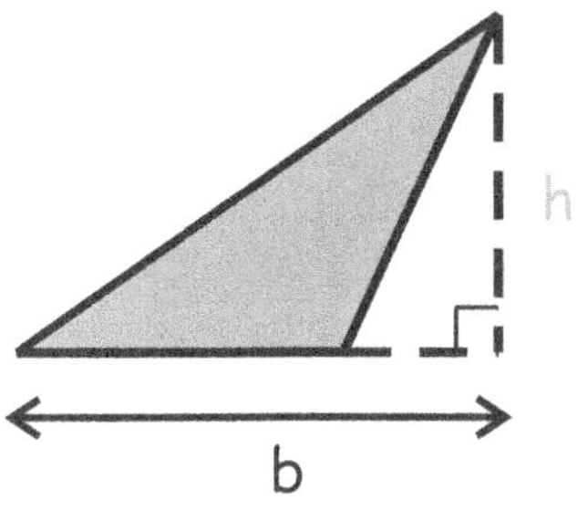

$A = \frac{1}{2} bh$

2. Area of a parellelogram

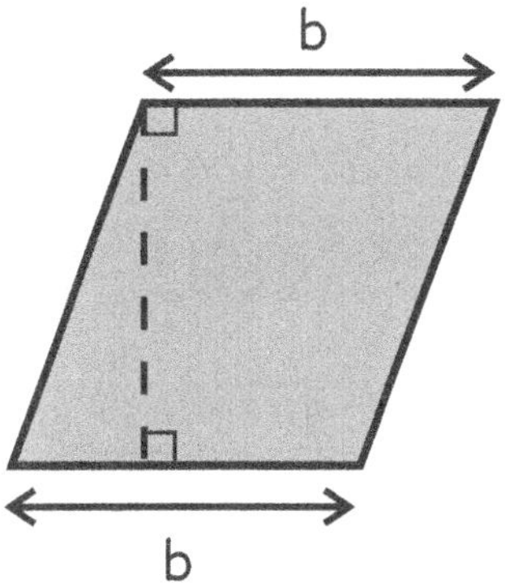

$A = bh$

3. Volume and Surface area of a cuboid

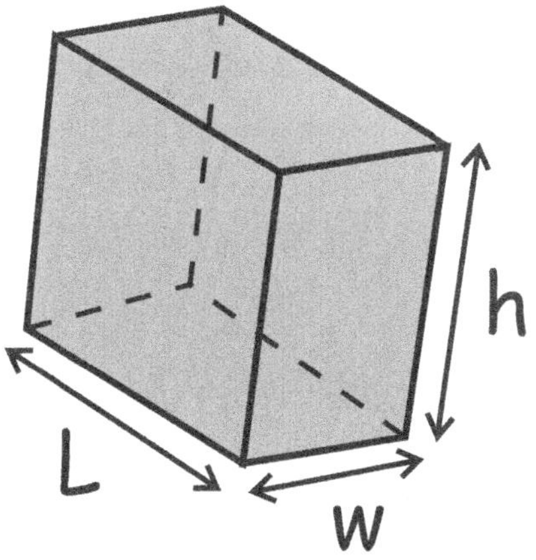

$V = lwh$

$S.A = 2(lw + lh + wh)$

4. Perimeter and Area of a Square

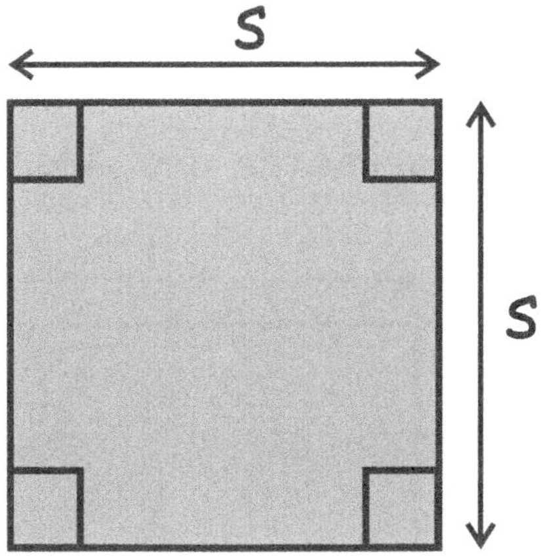

$P = 4s$

$A = s^2$

5. Area of a Trapezium

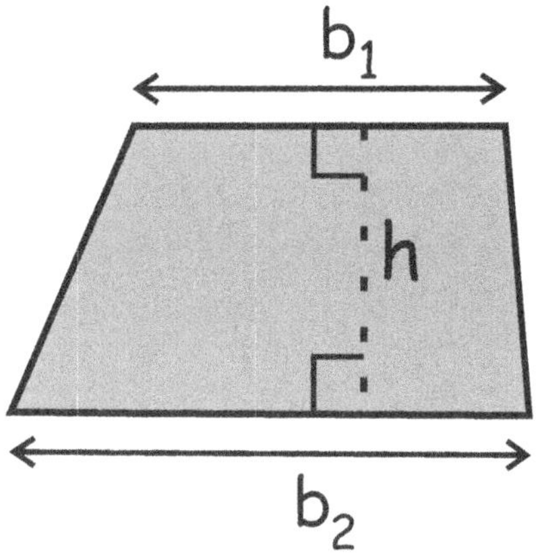

$$A = \frac{1}{2}h(b_1 + b_2)$$

6. Circumference and Area of a Circle

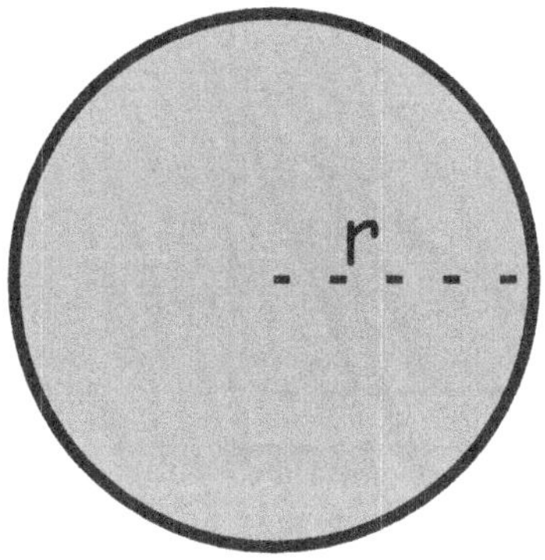

$$c = 2\pi r$$
$$A = \pi r^2$$

pi

$$\pi = 3.14$$
$$\pi = \frac{22}{7}$$

7. Perimeter and Area of a Rectangle

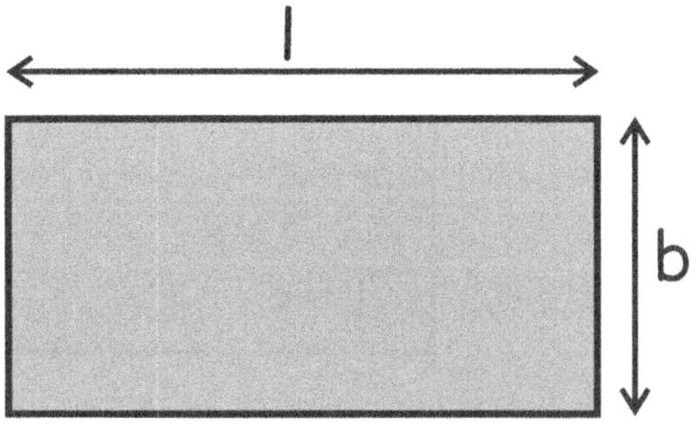

$\text{Area} = l \quad b \quad \times$

$\text{Perimeter} = 2(l + b)$

8. Acute angle

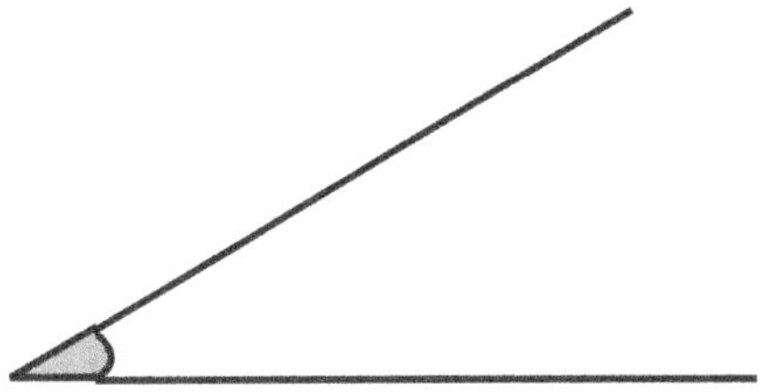

Any angle less than 90^0
is called as acute angled

9. Right angle

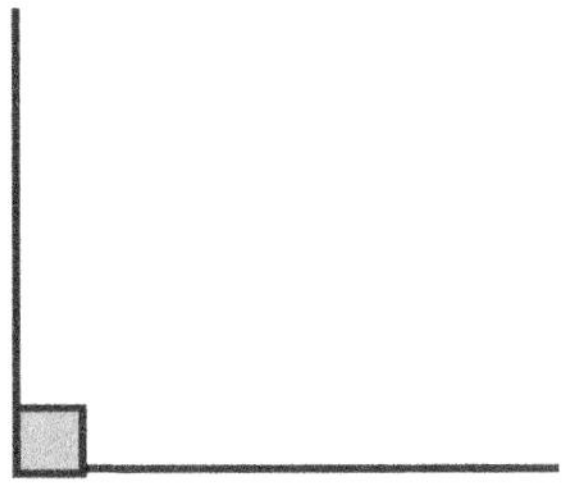

An angle equal to 90^0
is called as right angled

10. Obtuse angle

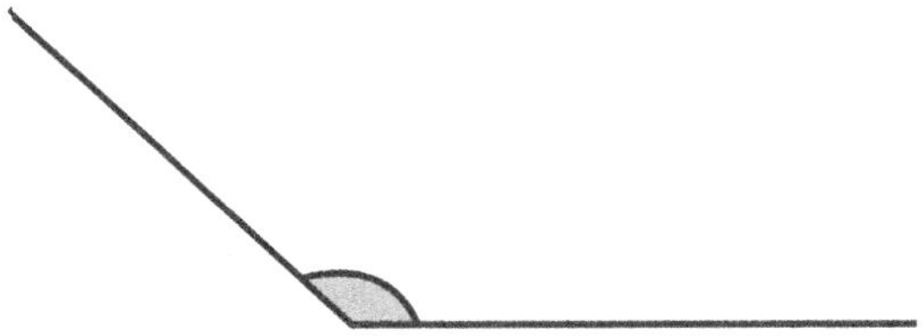

Any angle greater than 90^0
is called as obtuse angled

11. Perimeter of a polygon KLMNOP

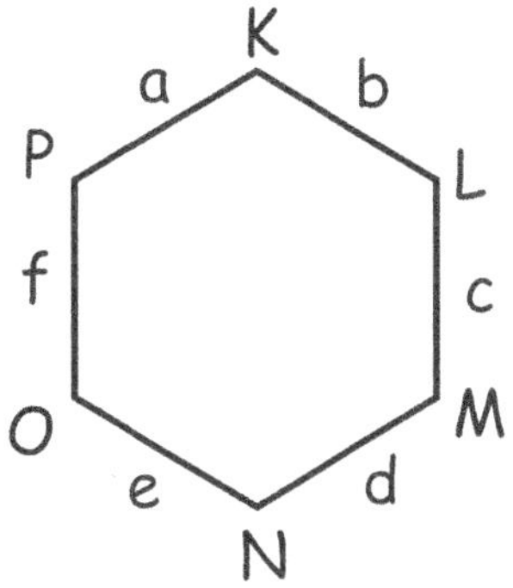

Perimeter =
(a + b + c + d + e + f) units

Abbreviations

milligram	mg
gram	g
kilogram	kg
milliliter	mL
liter	L
kiloliter	kL
millimeter	mm
centimeter	cm
meter	m
kilometer	km
square centimeter	cm^2
cubic centimeter	cm^3

year	yr
month	mon
hour	hr
minute	min
second	sec

volume	V
total Square Area	S.A
area of base	B
ounce	oz
pound	lb
quart	qt
gallon	gal.
inches	in.
foot	ft
yard	yd
mile	mi.
square inch	sq in.
square foot	sq ft
cubic inch	cu in.
cubic foot	cu ft

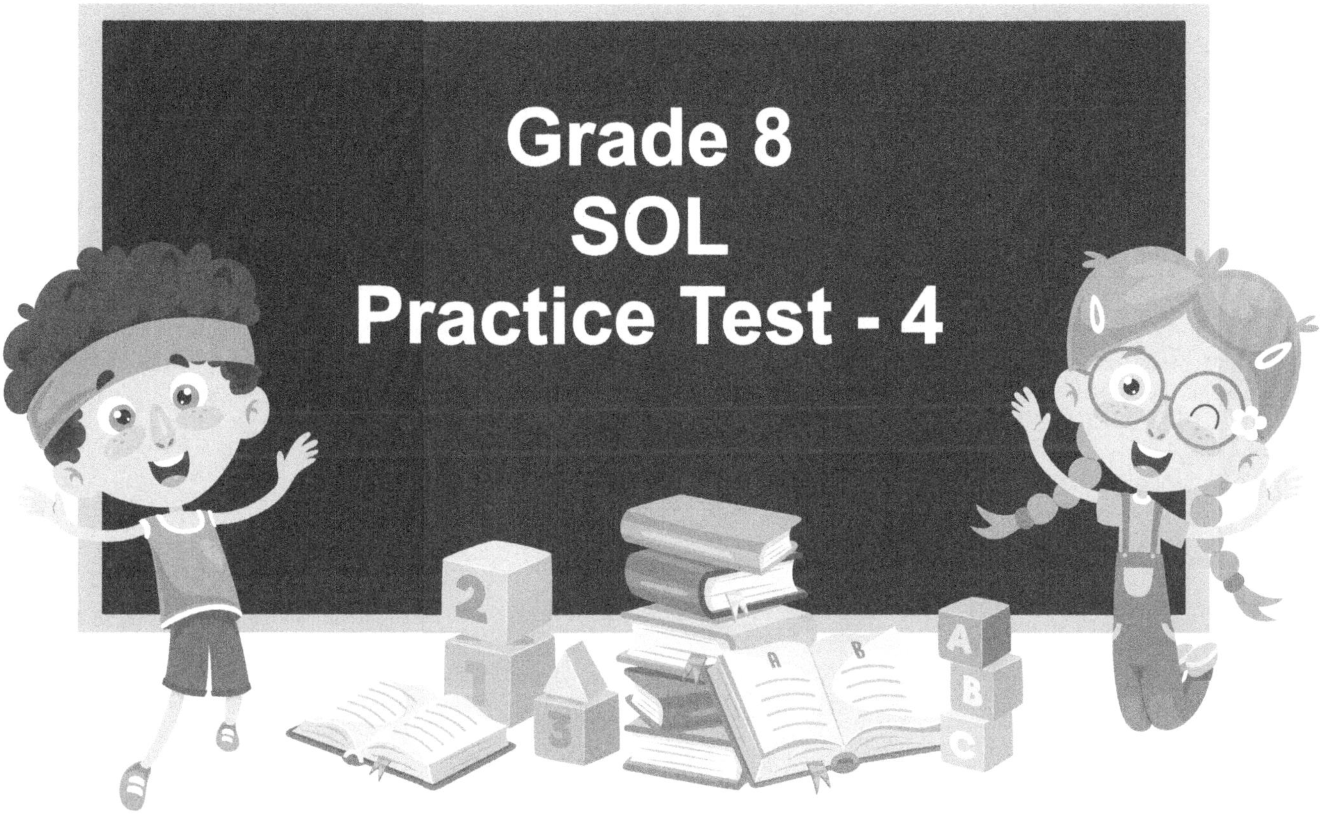
Grade 8
SOL
Practice Test - 4

1. Evaluate $(-3)((-18) - 3 - (-17))$

(A) 25 (B) 20

(C) 9 (D) 12

2. Simplify $4^2 - (13 - (-15))$

(A) -26 (B) -25

(C) -12 (D) -22

3. Which of the below is greater than 444444 ?

(A) 44.54×10^2 (B) 55.55×10^3

(C) 89.99×10^3 (D) 33.33×10^7

4. Which of the below is an irrational number ?

(A) $\sqrt{23}$ (B) -55

(C) $\frac{7}{15}$ (D) $\sqrt{4}$

5. Which of the below represents 0.000046 in scientific notation ?

(A) 0.46×10^{-2} (B) 0.46×10^{-5}

(C) $0.4.6 \times 10^{-5}$ (D) 0.46×10^{-3}

6. The set of rational number is a subset of ________

(A) Irrational number (B) Integers

(C) Whole numbers (D) Real numbers

7. Which of the following describes a square root of 39 ?

(A) Between 5 and 6 (B) Between 6 and 7

(C) Between 20 and 21 (D) Between 40 and 42

8. Simplify $m + n \times \frac{p}{6}$; where m = 6 , n = -6 and p = -6

(A) 2 (B) 6

(C) 12 (D) 16

9. Simplify $z\left(x - \left(\frac{5}{5}\right)^2\right)$; where x = 10 and z = -6

(A) -54 (B) -49

(C) -59 (D) -50

10. Which of the below is a perfect square ?

(A) 20 (B) 11

(C) 45 (D) 4

11. The square root of which of the following integers is between 15 and 16 ?

(A) 200 (B) 248

(C) 225 (D) 256

12. A mini van is on sale with a discount of 22% which is originally priced at $45,000.00. Find the selling price of mini van ?

(A) $35,100.00 (B) $51,750.00

(C) $49,500.00 (D) $9,900.00

13. Find the volume of a square prism measuring 19 yd along each edge of the base and 11 yd tall. Round the answer to the nearest tenth.

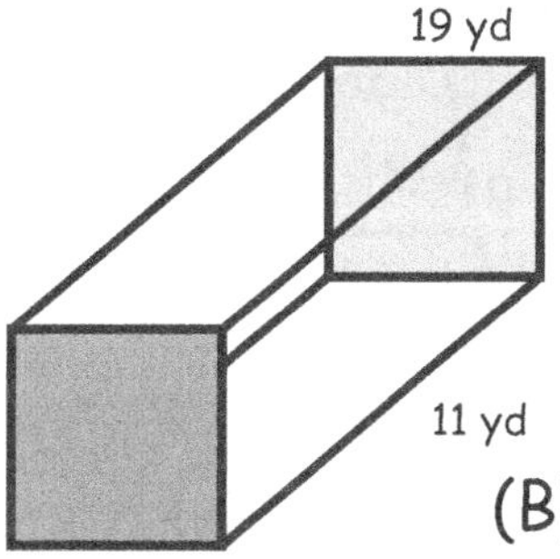

(A) 3971 yd^3 (B) 4447.5 yd^3

(C) 7827.6 yd^3 (D) 3913.8 yd^3

14. What is the relation between $\angle a$ and $\angle b$ from the below figure ?

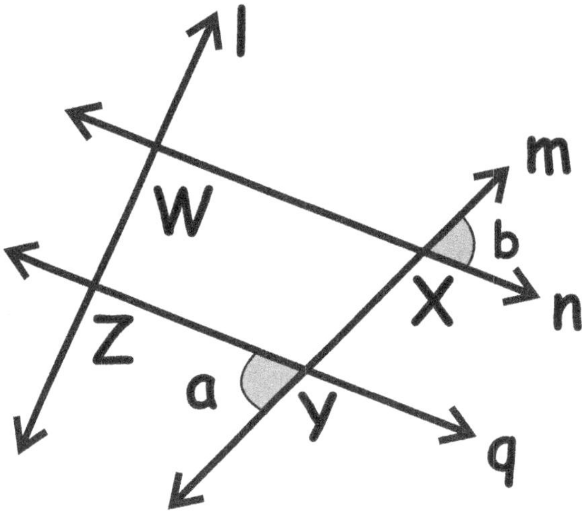

(A) Adjacent angles

(B) Complementary angles

(C) Alternate exterior angles

(D) alternate interior angles

15. Find the missing angle in the below figure.

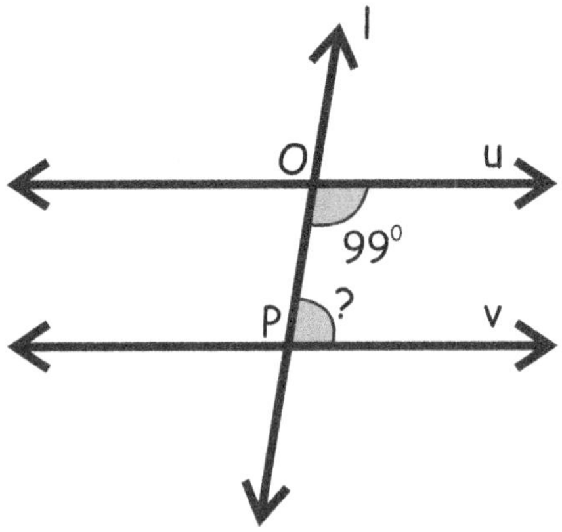

(A) 81^0

(B) 77^0

(C) 99^0

(D) 75^0

16. Which of the below options form a right angle triangle ?

(A) 60 , 63 , 77 (B) 24 , 32 , 40

(C) 40 , 42 , 52 (D) 12 , 35 , 36

17. Which of the below options describes the transformation of the below figure CED ?

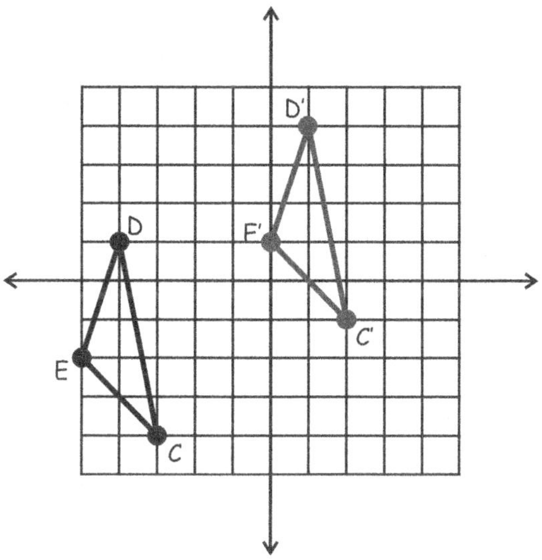

(A) translation of 6 units left and 4 units down

(B) rotation of 90^0 clockwise about the origin

(C) rotation of 90^0 counterclockwise about the origin

(D) translation of 5 units right and 3 units up

18. Find the volume of a cylinder with radius of 9 ft and a height of 18 ft. Answer to the nearest teth

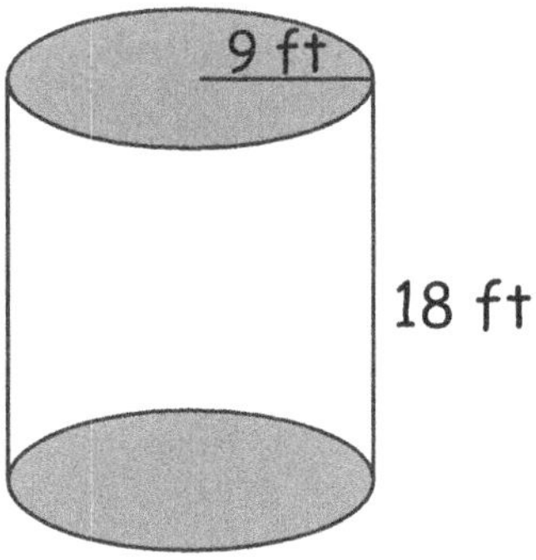

(A) 3478.3 ft^3

(B) 3664.4 ft^3

(C) 4580.4 ft^3

(D) 3078.1 ft^3

19. Which of the following equations is represented by the figure ?

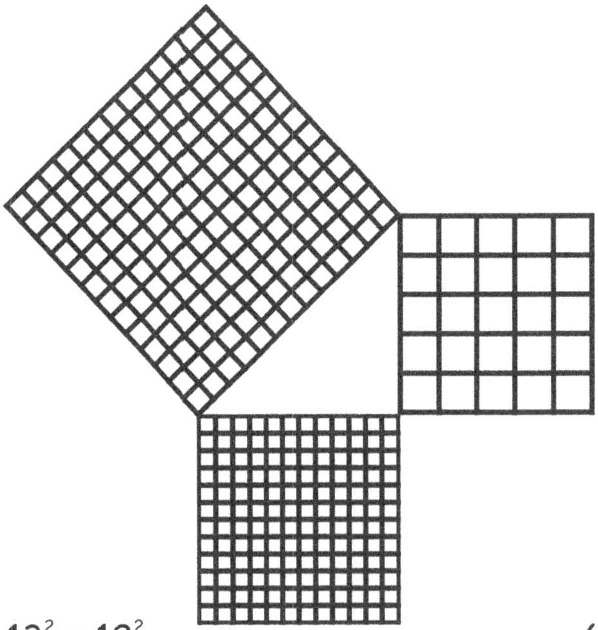

(A) $5^2 + 13^2 = 12^2$

(B) $13^2 + 12^2 = 5^2$

(C) $13^2 + 4^2 = 12^2$

(D) $12^2 + 5^2 = 13^2$

20. Tuck has a 6 sided dice marked by number 10 on five sides and number 1 is marked on the remaining side. Which statement best describes Tuck's chance of winning the game by getting the number 1 ?

(A) Certainly will win (B) Certainly will lose

(C) Most likely will win (D) Most likely will lose

21. Which of the below graphs represent the equation

$$y = x + 1$$

(A)

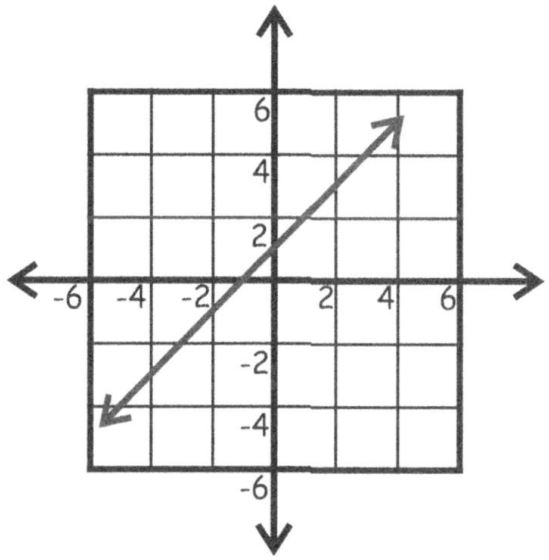

(B)

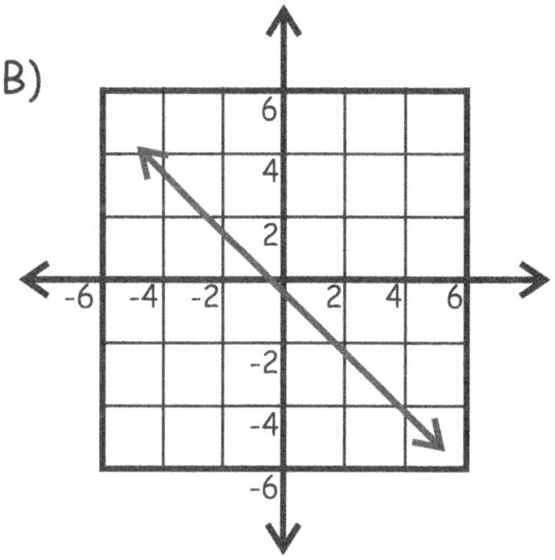

(C)

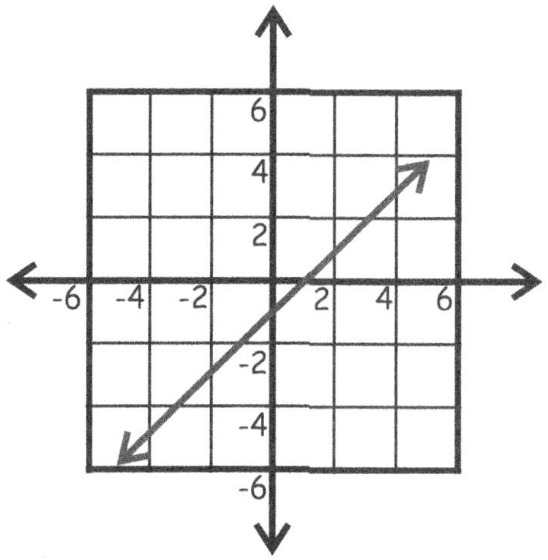

(D)

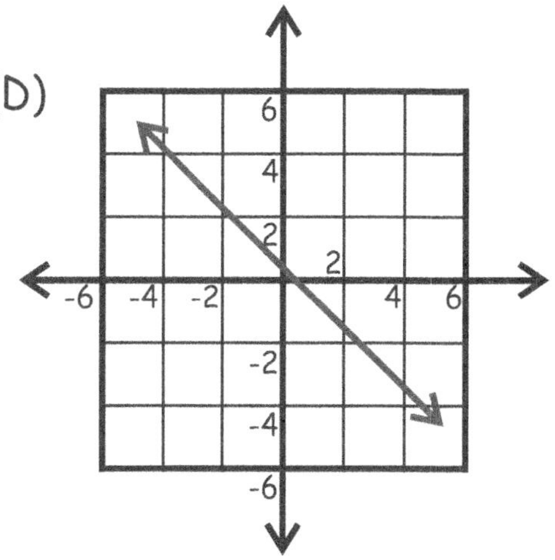

22. Simplify -21 >= -19 - x

(A) x >= $\frac{21}{19}$

(B) x >= 40

(C) x >= 2

(C) x >= -40

23. Find the value of n 75 = n - (-91)

(A) -16

(B) -87

(C) $-\frac{75}{91}$

(D) -71

24. Ryan's recipe of cup cakes needs $5\frac{3}{4}$ cups of brown sugar. He only has $5\frac{2}{3}$ cups of brown sugar at home. How many more cups of brown sugar does he need to bake the cup cakes ?

(A) $1\frac{1}{68}$

(B) $5\frac{5}{6}$

(C) $\frac{1}{12}$

(D) $11\frac{5}{12}$

25. The formula can be used to find the distance , d = 52t + 100 between two cars after traveling t hours in opposite directions. After how many hours will the distance between the two cars be 360 miles ?

(A) 5

(B) 0.5

(C) 3.8

(D) 4.7

26. James made a trip to the town hall and back. The trip there took five hours and the trip back took two hours. What was James average distance travelled in the trip if his average speed is 30 mph ?

(A) 145 miles (B) 150 miles

(C) 153 miles (D) 105 miles

27. Simplify n + (-15) <= -35

(A) n <= 20 (B) n <= -20

(C) n <= -50 (D) $n <= \frac{7}{3}$

28. What is the range of the data plotted ?

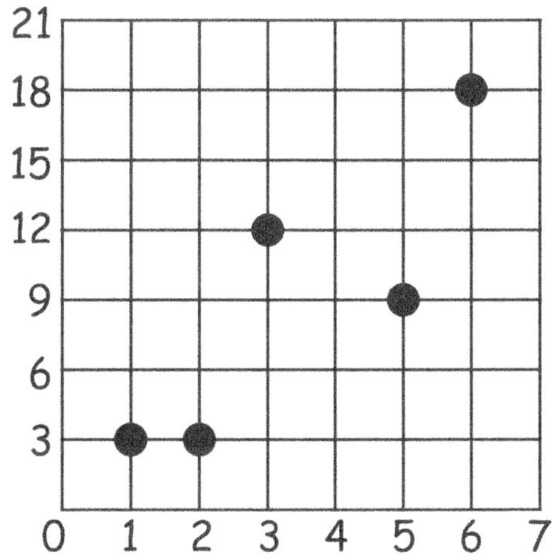

(A) {1,3,4,6} (B) {2,3,5,6}

(C) {3,3,9,12,18} (D) {0,1,2,3,4,5,6}

29. Stephan has a savings of $1,500 at the end of January,2019.He saves $100 every month after January,2019. Which graph correctly displays Stephan's savings ?

(A)

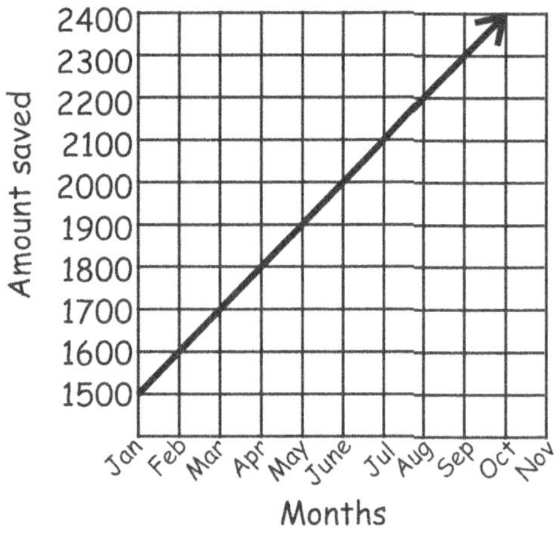

(B)

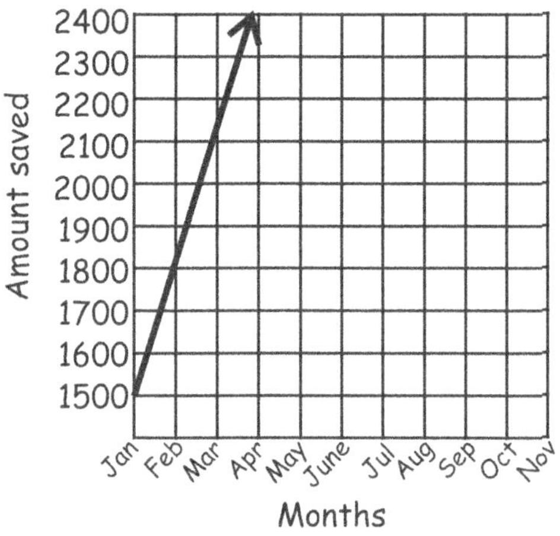

(C)

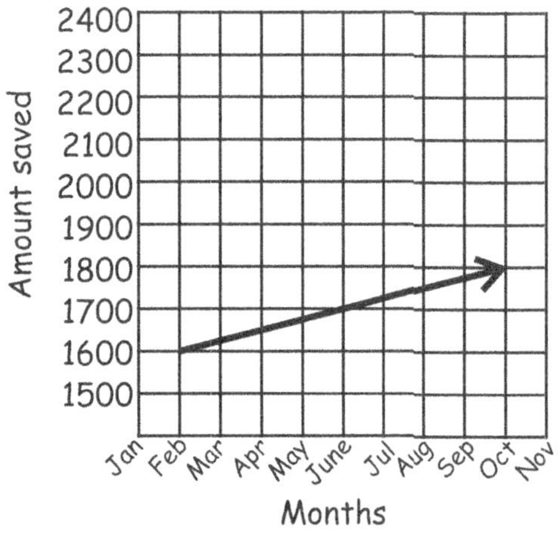

(D)

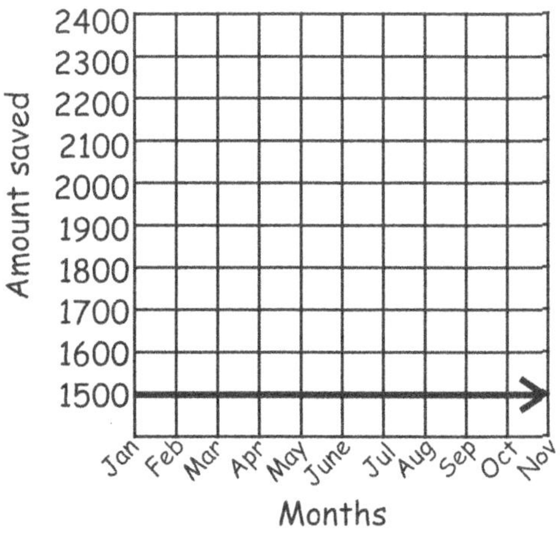

30. Which of the following tables of values is true for the equation

$$y = 2x ?$$

(A)

x	y
2	-5
-1	2
0	0
1	5

(B)

x	y
-2	-4
-1	-2
0	0
1	2

(C)

x	y
-2	-4
-5	-2
0	1
1	-3

(D)

x	y
-4	-3
-1	-2
0	5
1	-2

31. Which number is not an element of the domain of the relation shown?

{ (-13,5) , (0,7) , (9,12) , (-8,16) }

(A) -13

(B) -8

(C) 0

(D) -9

32. Which graph contains the ordered pairs listed in the table ?

x	0	-4	-2	2	4
y	0	-2	-1	1	2

(A)

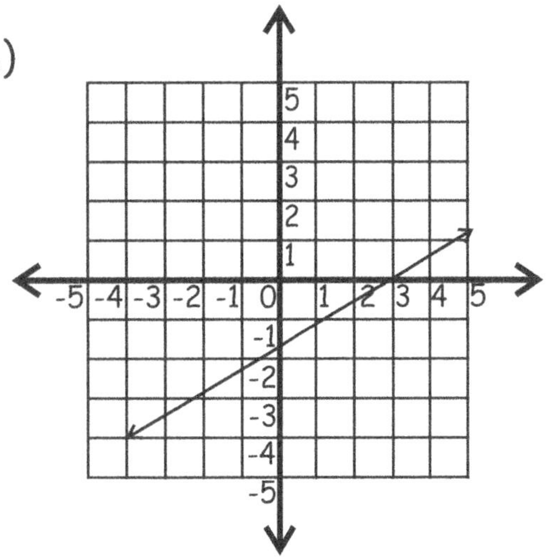

(B)

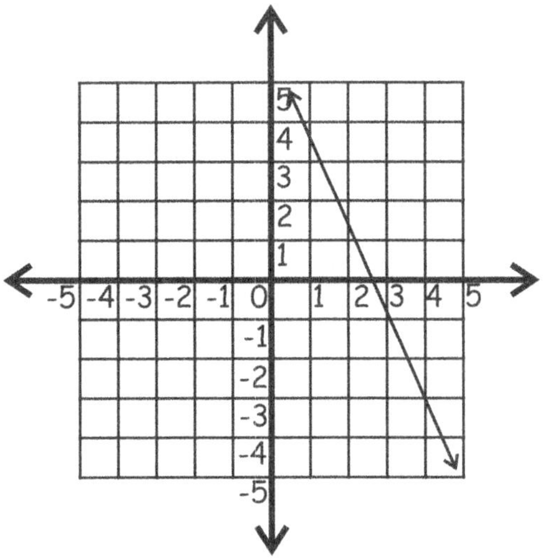

(C)

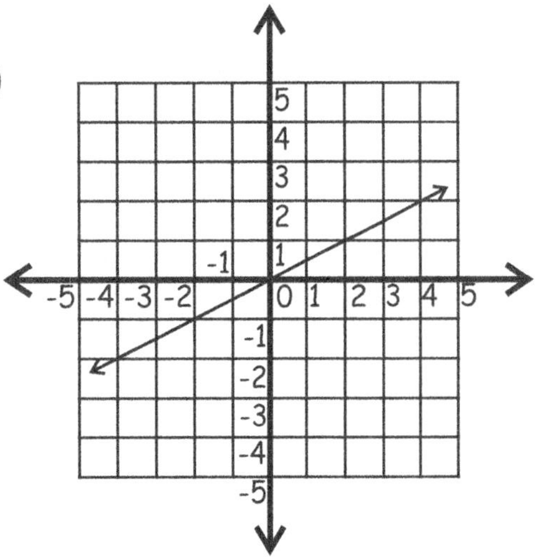

(D)

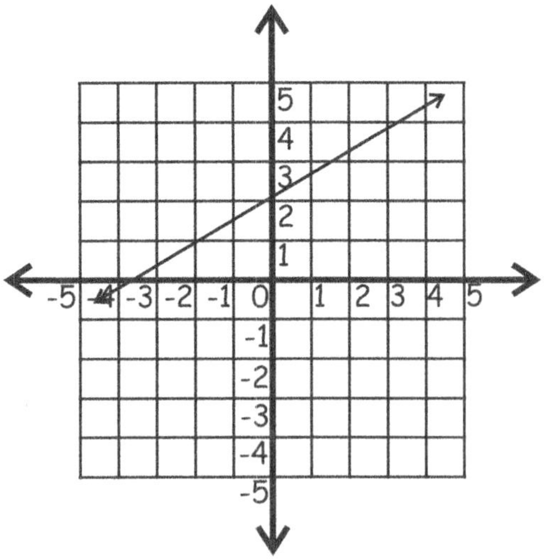

33. If the line containing the points in the table is plotted on a coordinate system , what does the graph look like ?

x	y
0	0
-2	1
-4	2
2	-1
4	-2

(A)

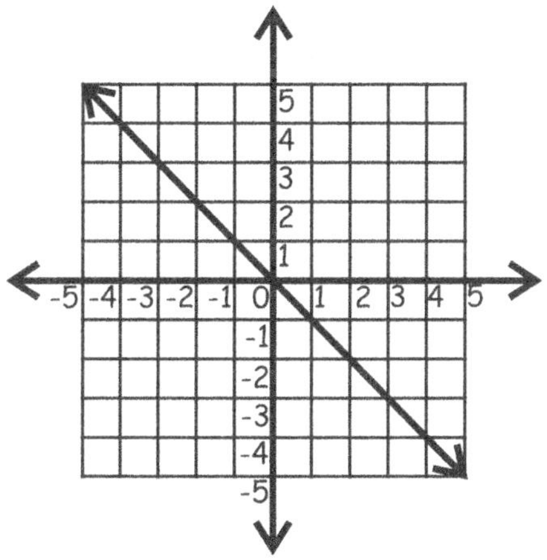

(B)

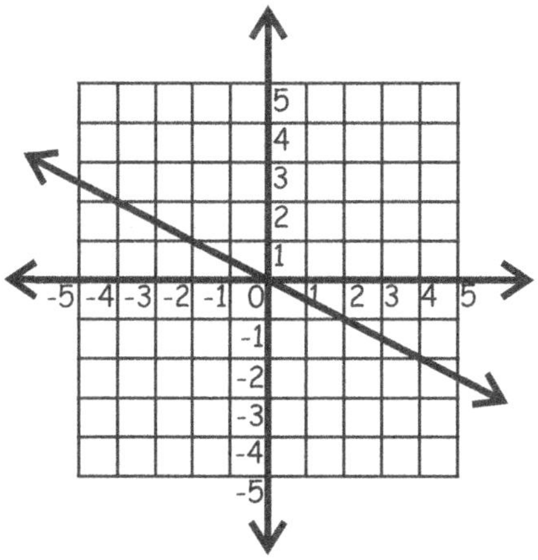

(C)

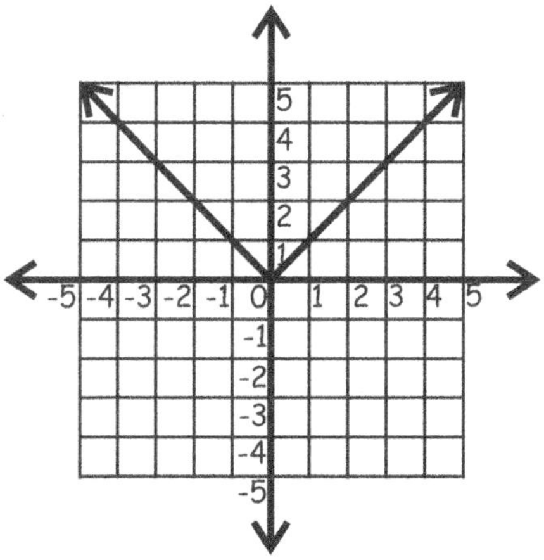

(D)

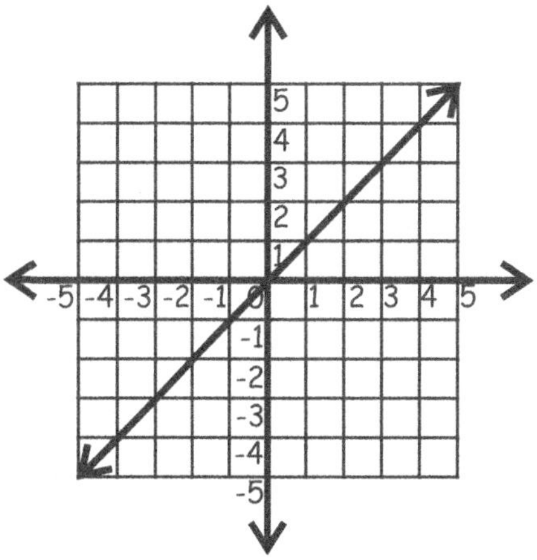

34. Which table contains only points that lie on the line of the equation

$$y = \frac{1}{3}x + 1$$

(A)

x	y
5	-1
3	2
6	3
7	4

(B)

x	y
7	-1
3	-2
6	-3
-9	4

(C)

x	y
-6	-1
3	2
6	3
9	4

(D)

x	y
-6	1
3	-2
0	3
-2	5

35. Which of these equations represents the table of values ?

x	-1	0	1	10
y	-10	-9	-8	1

(A) $y = x - 9$

(B) $y = 2x + 9$

(C) $y = 3x - 9$

(D) $y = x + 9$

36. The below box and whisker plots summerize the social studies test scores of grade 8 and grade 7

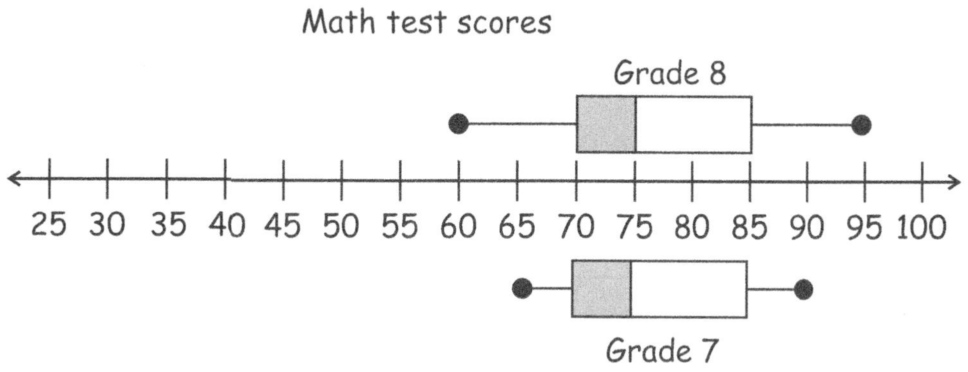

Based on the above data , which of the below statements is true ?

(A) The least score of grade 7 students is 55
(B) No grade 7 students got a score of 75
(C) The least score of grade 8 students is 40
(D) The highest score of grade 8 students is 95

37. The matrix below shows average class scores of grade 6 , grade 7 and grade 8 from 3 sections A , B and C in math midterm test. Which element represents the value of row 1 , column 2 ?

	Gr - 6	Gr - 7	Gr - 8
Sec - A	75	77	70
Sec - B	67	95	99
Sec - C	84	92	95

(A) 67
(B) 99
(C) 77
(D) 84

38. The below scatter plot describes the winning scores of soccer team and the number of weeks of practice.

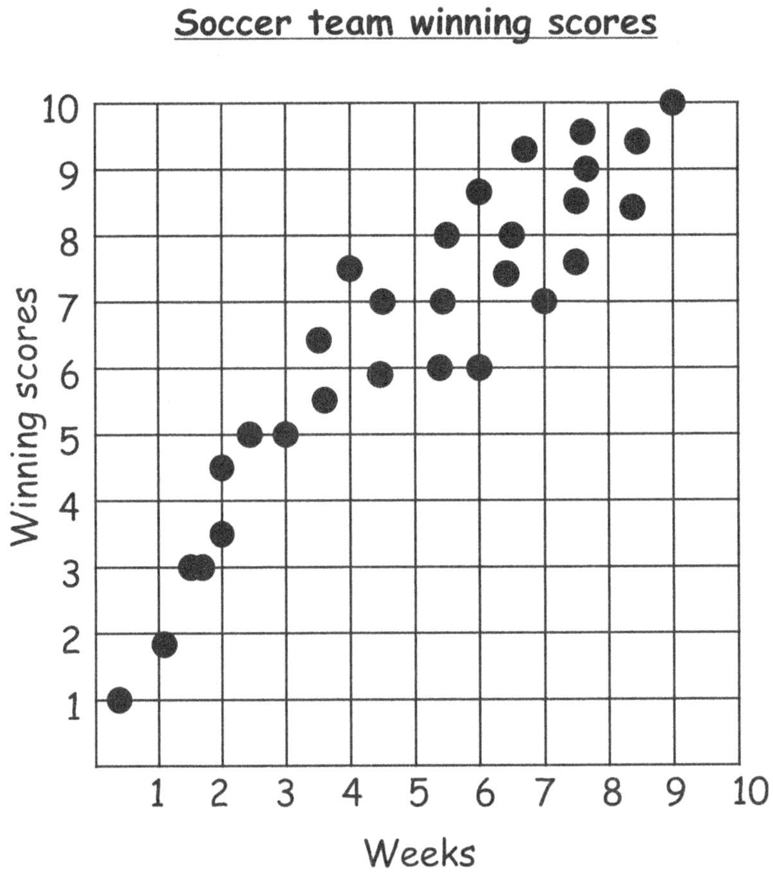

Based on the above data , which statement best describes the relationship between team's practice and win's ?

(A) Less practice gave more wins to the team
(B) There is no correlation between practice and high scores
(C) More weeks of practice resulted in winning the games
(D) More practice made the team lose the games

39. A box contains 12 squares , 8 cubes , 9 triangles , 11 prisms , 10 cones. Danny picks an item from the box , what is the probability that it will be a cube or a square ?

(A) 44 %
(B) 40 %
(C) 58 %
(D) 49 %

40. Three friends share riddles with each other as given below.

- Jenna shares 11 riddles with Laura and 9 riddles with Alison
- Laura shares 8 riddles with Alison and 6 riddles with jenna
- Alison shares 12 riddles with Jenna and 9 riddles with Laura

Which of the below matrix summarizes the above information ?

(A)

	Jenna	Laura	Alison
Jenna	1	11	9
Laura	8	0	5
Alison	10	9	4

(B)

	Jenna	Laura	Alison
Jenna	0	10	9
Laura	8	7	6
Alison	15	9	0

(C)

	Jenna	Laura	Alison
Jenna	0	11	9
Laura	6	0	8
Alison	12	9	0

(D)

	Jenna	Laura	Alison
Jenna	0	12	8
Laura	9	0	6
Alison	11	9	0

41. Lucy has $2000 in his savings account. Her mom deposits $75 every week. Assuming Lucy makes no withdrawals from the account. Which table represents her account statement ? (Excluding any interest amount associated)

(A)

Week	Account balance
0	$2075
1	$2100
2	$2150
3	$2225
4	$2300

(B)

Week	Account balance
0	$2000
1	$2075
2	$2150
3	$2225
4	$2300

(C)

Week	Account balance
0	$2000
1	$2075
2	$2175
3	$2250
4	$2300

(D)

Week	Account balance
0	$2000
1	$2075
2	$2150
3	$2250
4	$2325

42. This graph shows the number of sandwiches sold each day during one week in the super sandwich store.

Day of Week	Number of Sandwiches sold
Monday	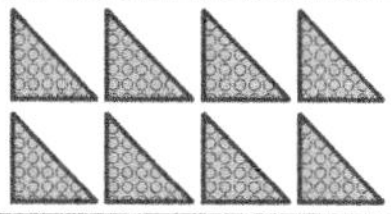
Tuesday	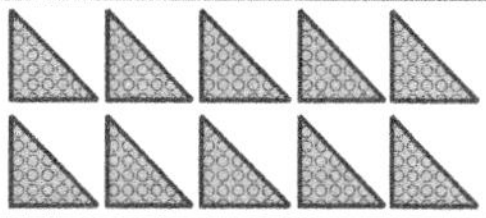
Wednesday	
Thursday	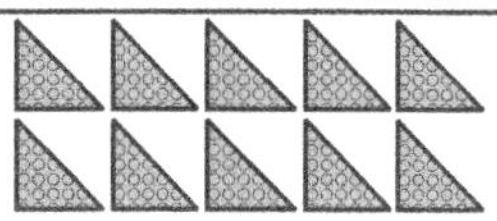
Friday	

Each sandwich costs \$3.00. What is the sale amount of the week ?

(A) \$155

(B) \$146

(C) \$162

(D) \$165

The bar graph given below shows the sales of books (in thousand number) from six branches of a publishing company during two consecutive years 2018 and 2019. Answer the question 43-45.

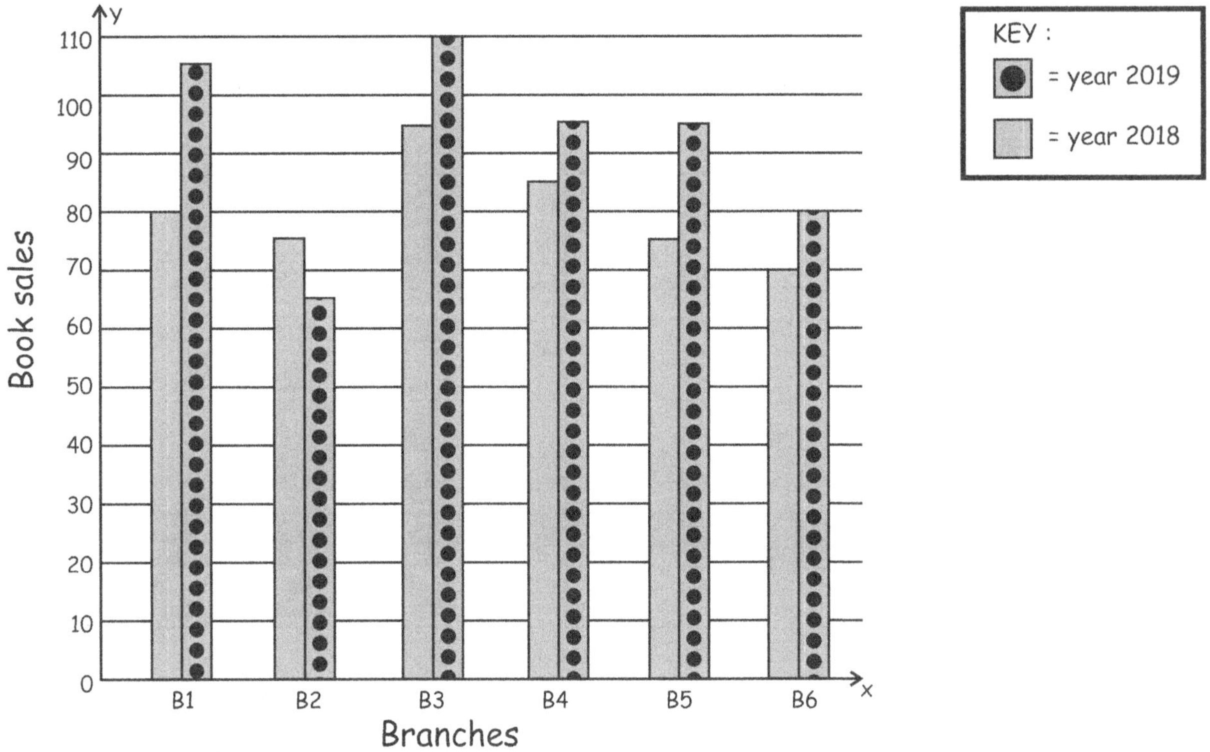

43. What is the ratio of the total sales of branch B2 for both years to the total sales of branch B4 for both years ?

(A) 2:3 (B) 3:5

(C) 4:5 (D) 7:9

44. Find the approximate average sales of branches B1 , B3 , B5 in the year 2019.

(A) 103 (B) 105

(C) 109 (D) 124

45. Find the approximate average sales of branches B2 , B3 , B4 in the year 2018.

(A) 80 (B) 85

(C) 125 (D) 75

46. Evaluate $(3 \times 10^{-2})^3 = ?$

(A) 3×10^{-2} (B) 3×10^{-4}

(C) 2.7×10^{-4} (D) 2.7×10^{-5}

47. Evaluate $\dfrac{5.3 \times 10^{-1}}{4 \times 10^{-5}} = ?$

(A) 1.325×10^{-4} (B) 1.325×10^{2}

(C) 1.325×10^{4} (D) 1.325×10^{1}

48. Evaluate $(2 \times 10^{-2})(9.71 \times 10^{0}) = ?$

(A) 19.42×10^{-4} (B) 1.942×10^{-1}

(C) 19.42×10^{-1} (D) 2.06×10^{-3}

49. The radius of a circular path is 15 feet. A 2 feet wide path is laid inside along its circumference. Find the area of the circular path.

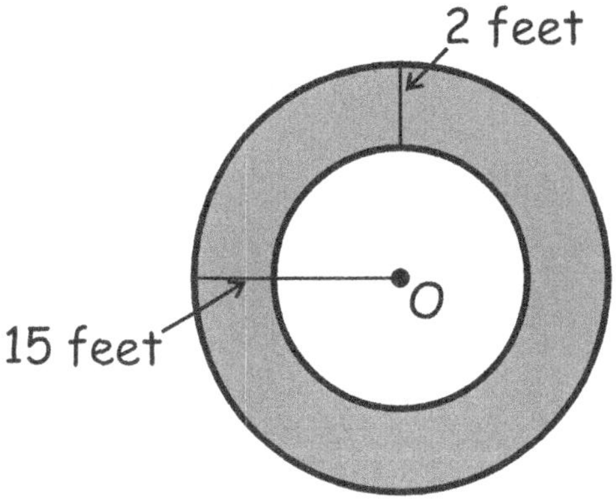

(A) 176 ft^2 (B) 179 ft^2

(C) 184 ft^2 (D) 172 ft^2

50. A wire in the form of a square is of length 18.84 cm on each side. The wire is reshaped to form a circle. What is the radius of the circle (Note : ∏ = 3.14)

(A) 15 cm (B) 11 cm

(C) 18 cm (D) 12 cm

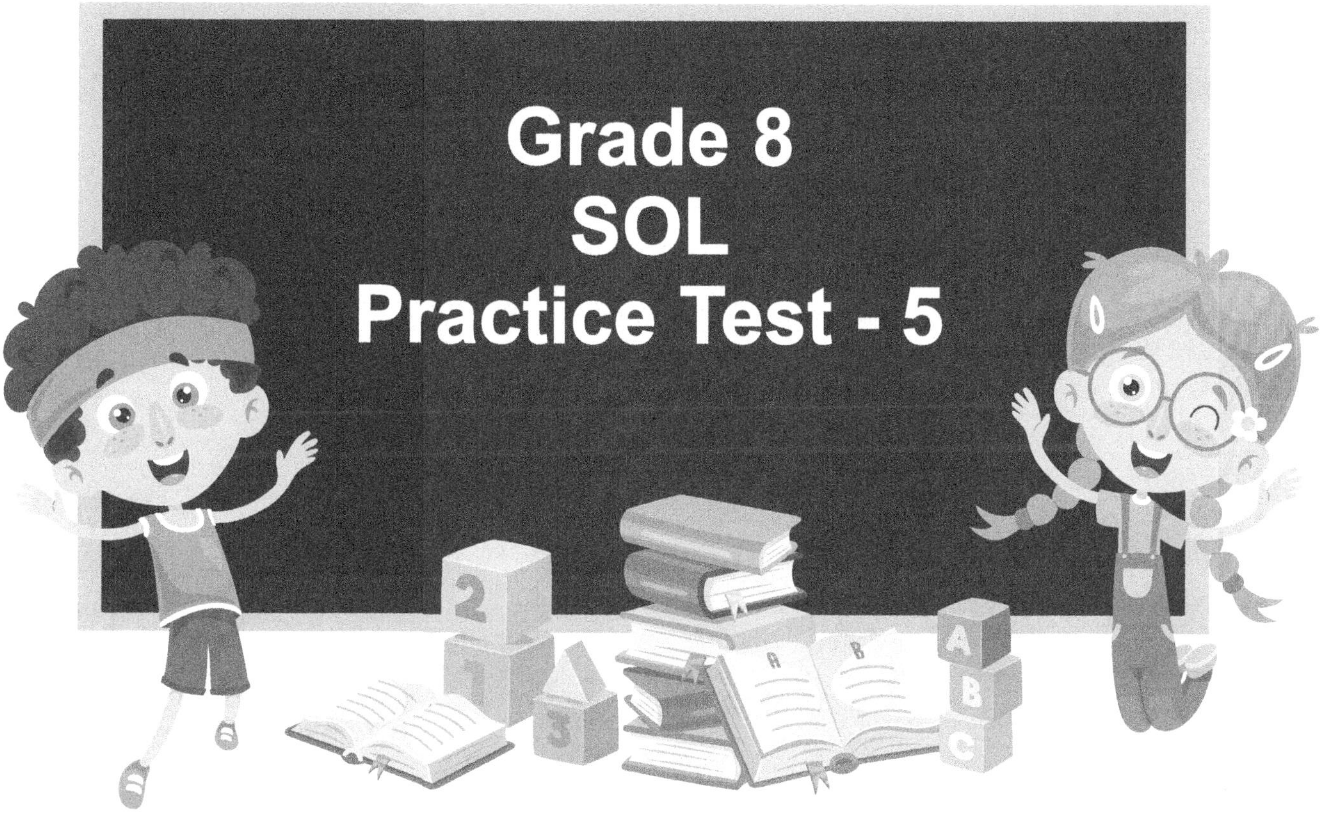
Grade 8
SOL
Practice Test - 5

1. Evaluate $((-3)-1) \div (-1)^3$

(A) 4 (B) 14

(C) 24 (D) -6

2. Simplify $14 + (-9)^2 + 1$

(A) 102 (B) 96

(C) 115 (D) 94

3. Which of the below is greater than 7676758 ?

(A) 63.125×10^6 (B) 63.12×10^0

(C) 89.12×10^3 (D) 65.55×10^{-5}

4. Which of the below is not an irrational number ?

(A) $\sqrt{23}$ (B) $\sqrt{5}$

(C) $\sqrt{89}$ (D) $\sqrt{49}$

5. Which of the below represents 7.93 in scientific notation ?

(A) 7.93×10^0 (B) 7.93×10^1

(C) 79.3×10^1 (D) 0.793×10^0

6. The set of whole numbers is a subset of ________

(A) Rational numbers (B) Integers

(C) Real numbers (D) Natural numbers

7. Which of the following describes a square root of 51 ?

(A) Between 25 and 26 (B) Between 8 and 9

(C) Between 7 and 8 (D) Between 49 and 52

8. Simplify $x + y - yz$; where $x = -2$ and $y = 8$ and $z = -3$

(A) 30 (B) 27

(C) 32 (D) 26

9. Simplify $z^3 + yx + z$;where $x = -9$ and $y = -6$ and $z = 4$

(A) 113 (B) 122

(C) 129 (D) 115

10. Which of the below is a perfect square ?

(A) 90 (B) 220

(C) 121 (D) 114

11. The square root of which of the following integers is between 5 and 6 ?

(A) 25 (B) 36

(C) 41 (D) 30

12. Owen purchased a digital camera at $140.00 and sells it at a 80% price increase. Find the selling price of the digital camera.

(A) $252 (B) $168

(C) $1112 (D) $126

13. Find the volume of a rectangular prism measuring 11 and 5 in along the base and 2 in tall. Round the answer to the nearest tenth.

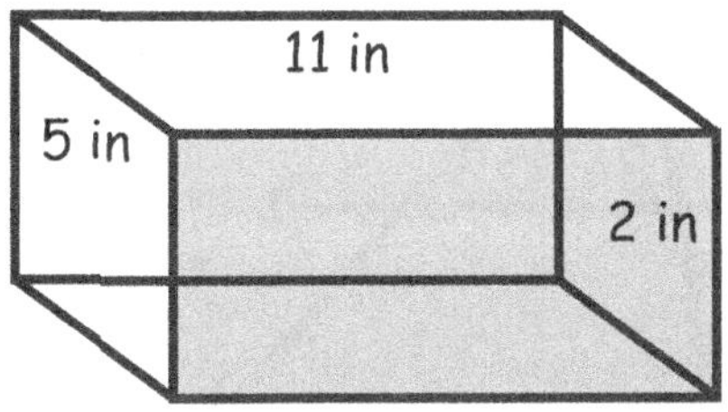

(A) 62.7 in^3 (B) 55 in^3

(C) 110 in^3 (D) 131.6 in^3

14. What is the relation between $\angle a$ and $\angle b$ from the below figure ?

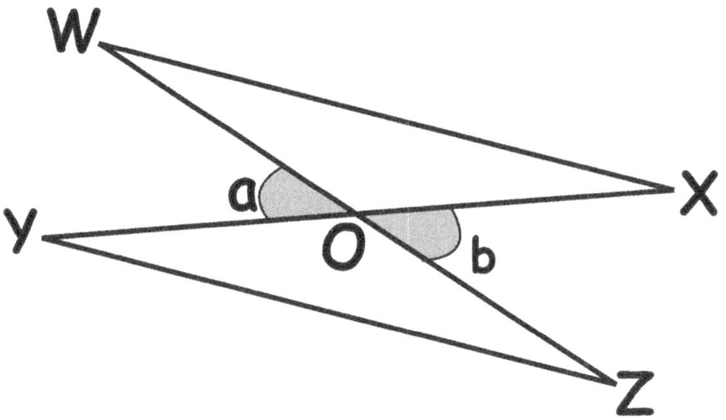

(A) Vertical angles

(B) Complementary angles

(C) Alternate exterior angles

(D) alternate interior angles

15. Find the missing angle in the below figure.

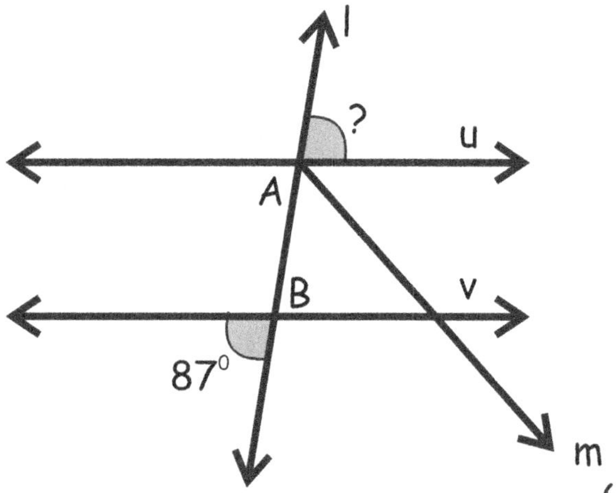

(A) 65^0

(B) 79^0

(C) 13^0

(D) 87^0

16. Which of the below options form a right angle triangle.

(A) 33 , 56 , 62 (B) 39 , 80 , 84

(C) 28 , 45 , 53 (D) 28 , 45 , 47

17. Which of the below options describes the transformation of the below figure EFGH ?

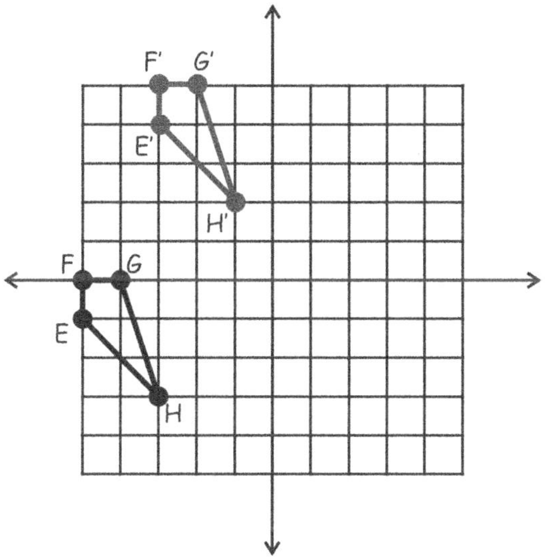

(A) translation of 2 units right and 5 units up

(B) reflection across x = -5

(C) rotation of 90° counterclockwise about the origin

(D) translation of 5 units left and 2 units down

18. Find the volume of a cylinder with radius of 1 cm and a height of 1 cm. Answer to the nearest teth

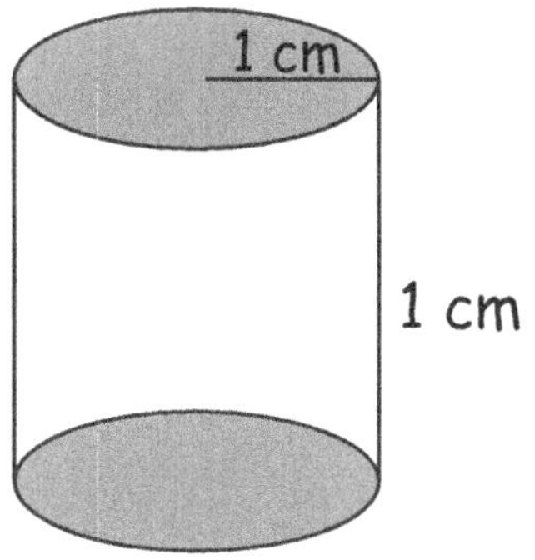

(A) $2.5\ cm^3$ (B) $5\ cm^3$

(C) $2.1\ cm^3$ (D) $3.1\ cm^3$

19. Which of the below is the hypotenuse ?

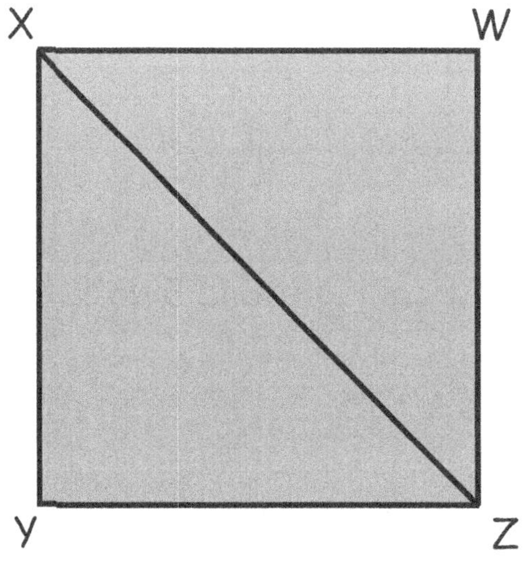

(A) XY (B) YZ

(C) XZ (D) None

20. Beth has a 6 sided dice. marked by number 6 on all sides. Which statement best describes Beth's chance of getting a 6 and win the game?

(A) Certainly will win

(B) Certainly will lose

(C) Most likely will win

(D) Most likely will lose

21. Which of the below graphs represent the equation

$$y = x - 5$$

(A)

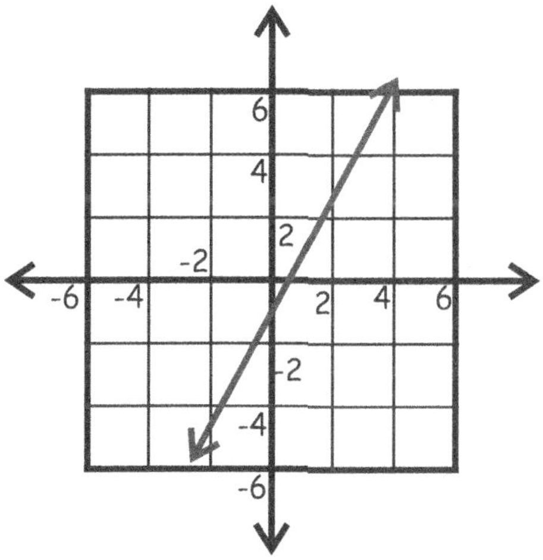

(B)

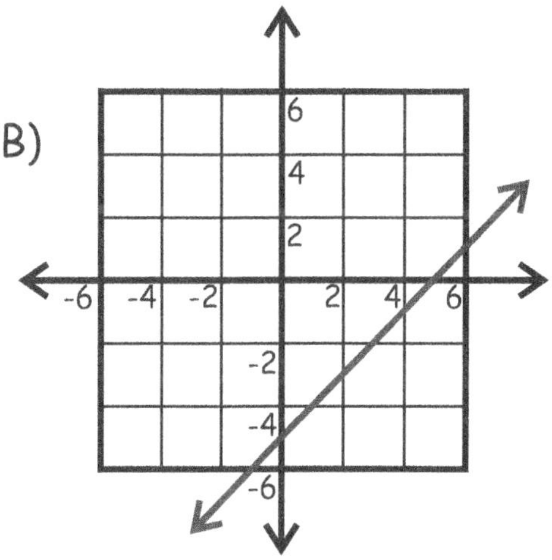

(C)

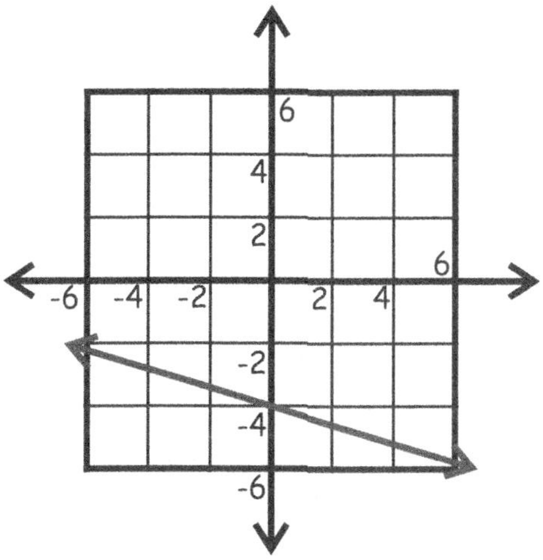

(D)

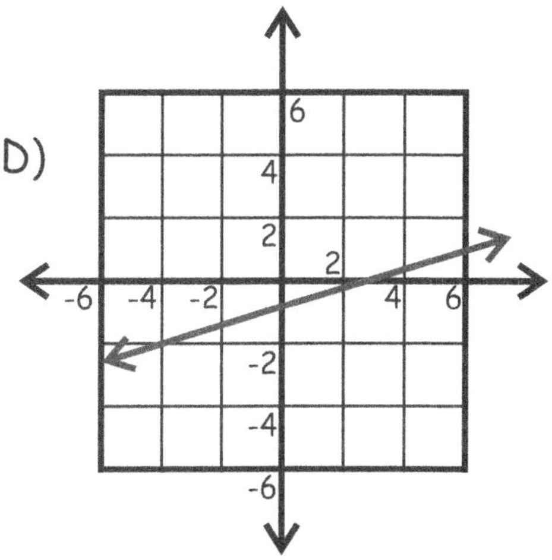

22. Simplify $-10 - x > 1$

(A) $x < -10$ (B) $x < -11$

(C) $x > -10$ (C) $x > -11$

23. Find the value of b

$$b - (-14) = 6$$

(A) -8 (B) -84

(C) $-\frac{3}{7}$ (D) 20

24. The weight of a carton of apples multiplied by $\frac{5}{6}$ gives 82 pounds. Find the weight of the carton of apples.

(A) 104.3 (B) 117.9

(C) 104.9 (D) 98.4

25. Fran drove m miles in a car she rented. To find R, the total cost of renting the car, she used the following formula.

$$R = 20m + 70$$

What is the dependent variable in the formula ?

(A) R (B) 0.50

(C) m (D) 35"

26. Working alone , it takes Amanda eight hours to paint a fence. Jacob can paint the same fence in nine hours. How long would it take them if they worked together ?

(A) 5.99 hours (B) 7.48 hours

(C) 5.66 hours (D) 4.24 hours

27. Simplify: $-\frac{7}{3} > \frac{p}{3}$

(A) $p < -7$ (B) $p > -\frac{16}{3}$

(C) $p < -\frac{16}{3}$ (D) $p > \frac{7}{9}$

28. What is the domain of this region ?

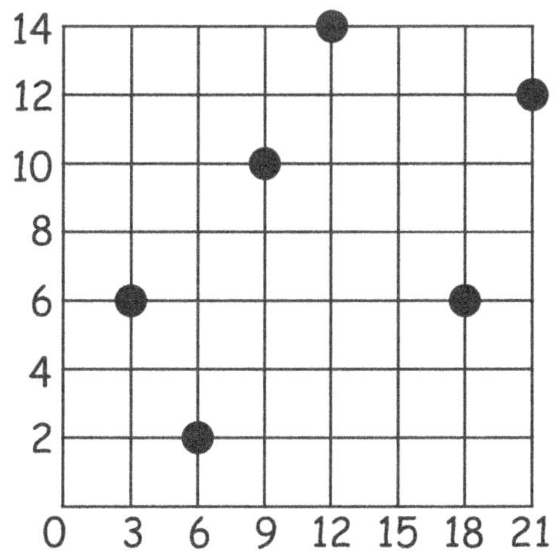

(A) {3,6,9,11,18} (B) {3,6,9,12,18,21}

(C) {2,3,4,5,6,12,18,21} (D) {0,4,5,6,7,11,18,19,21}

29. John gets a weekly pay of \$75 plus \$25 for every sale he does at the store. Which graph correctly displays John's earning across the sales ?

(A)

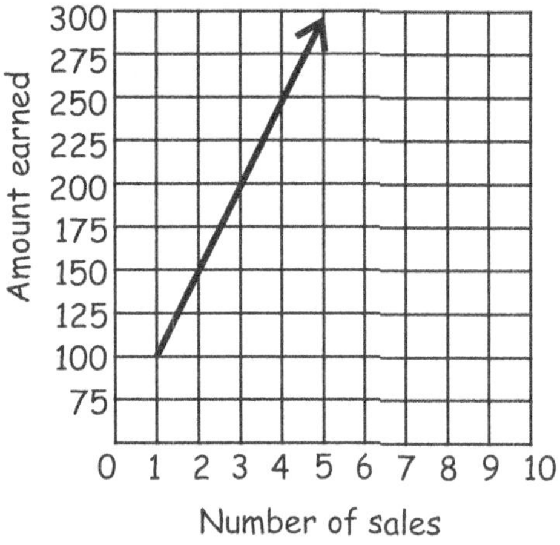

(B)

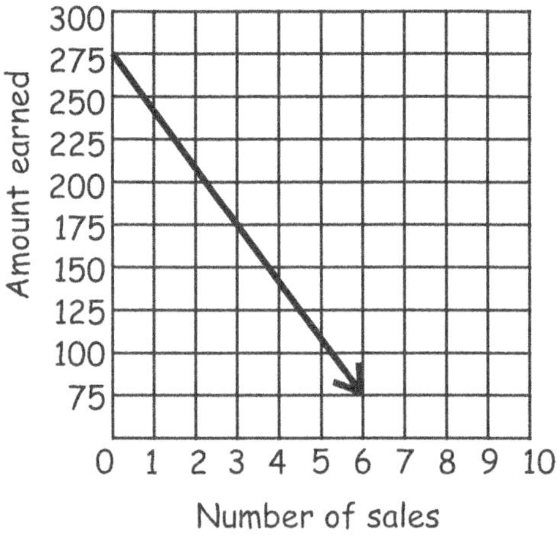

(C)

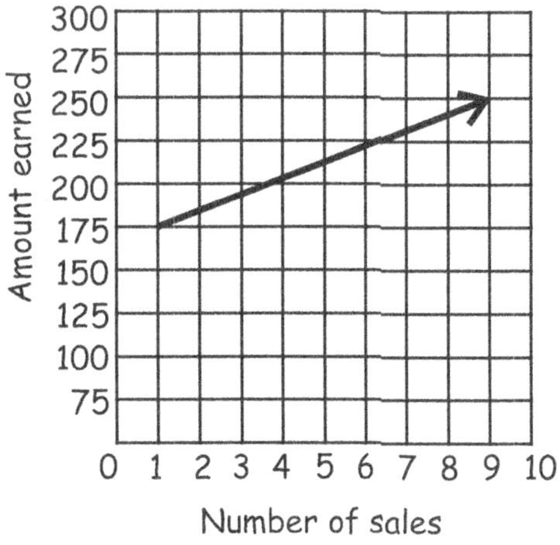

(D)

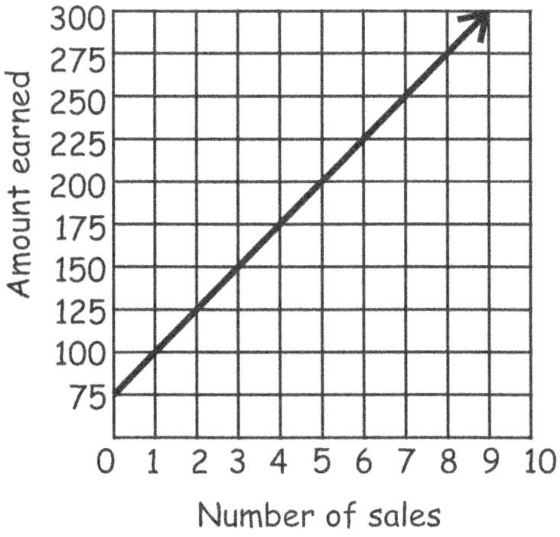

30. Which of the following tables of values is true for the equation

$$y = 7x - 1$$

(A)

x	y
-2	-10
2	13
0	1
3	20

(B)

x	y
-2	-15
2	12
0	5
7	20

(C)

x	y
-2	-15
2	13
0	-1
3	20

(D)

x	y
-2	-15
7	13
0	-1
3	25

31. Which number is not an element of the domain of the relation shown ?

{ (-11,11) , (-10,10) , (-7,6) , (-9,9) , (6,5) }

(A) 7
(B) -9
(C) 6
(D) -11

32. Which graph contains the ordered pairs listed in the table ?

x	-5	-3	2	4
y	-5	-3	2	4

(A)

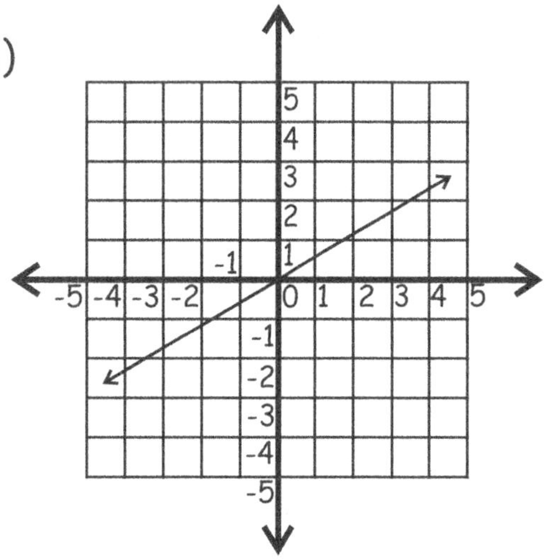

(B)

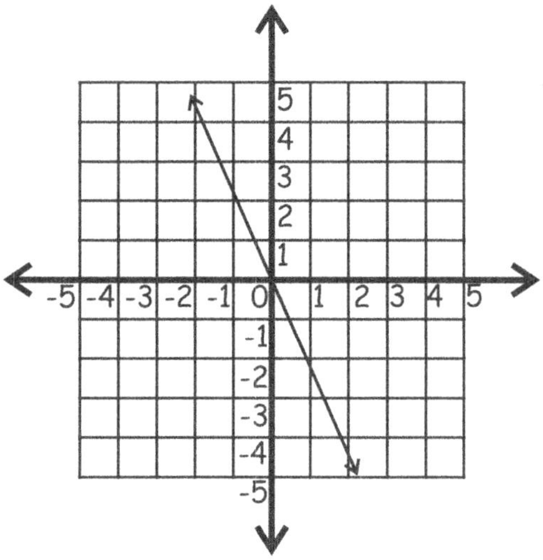

(C)

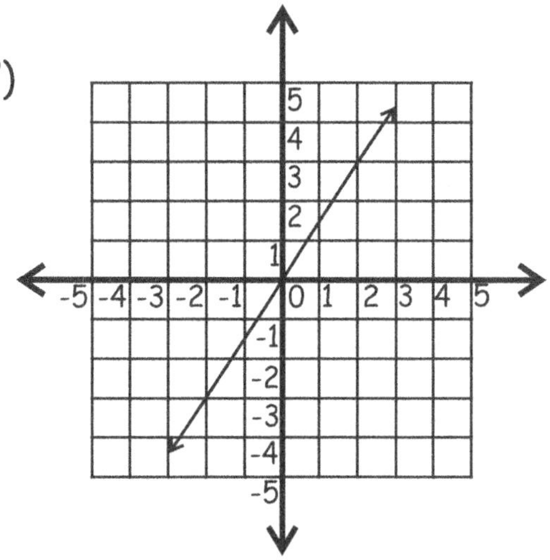

(D)

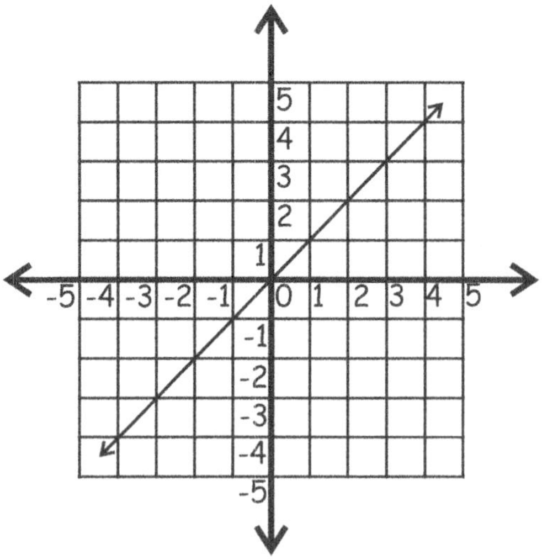

33. If the line containing the points in the table is plotted on a coordinate system,what does the graph look like ?

x	y
1	7
2	10
-1	1
-2	-2
-3	-5

(A)

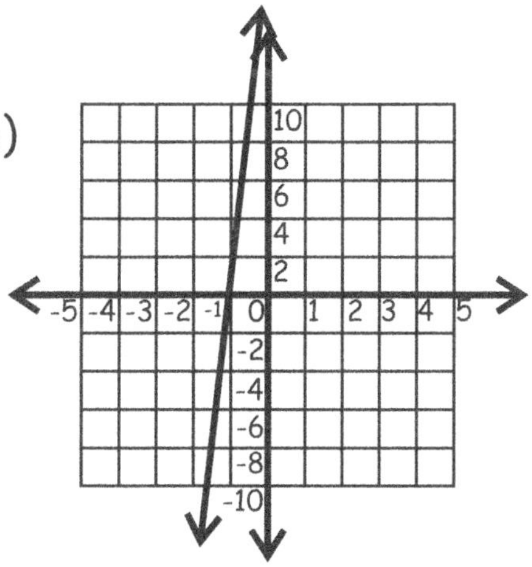

(B)

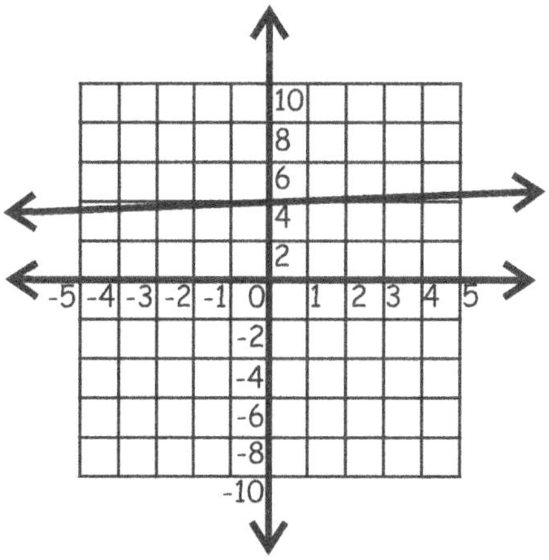

(C)

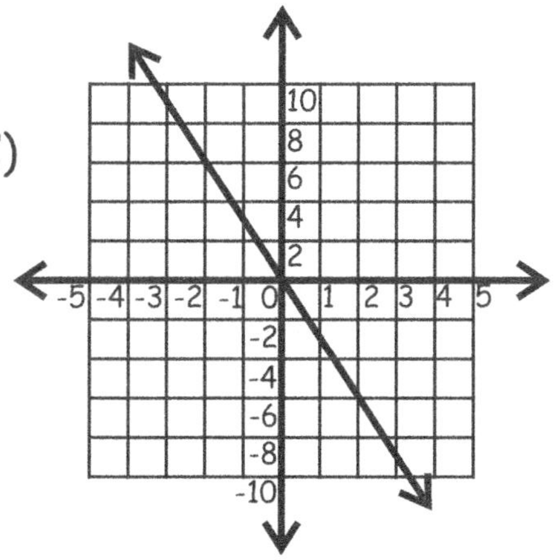

(D)

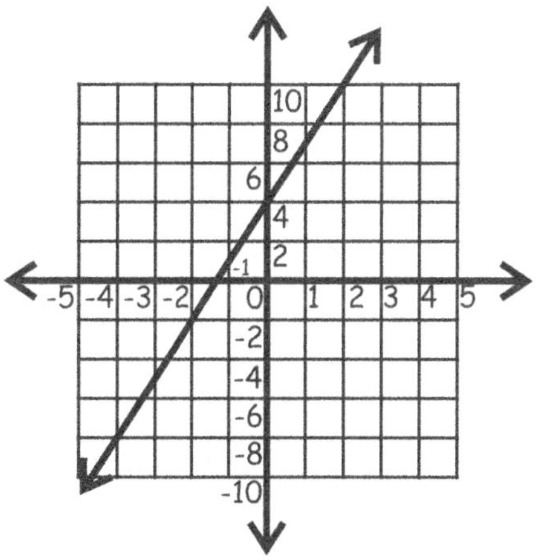

34. Which table contains only points that lie on the line of the equation $y = 2x - 2$?

(A)

x	y
0	-2
1	0
2	2
-1	-4

(B)

x	y
0	-2
1	0
2	-2
3	-4

(C)

x	y
-3	-2
1	0
3	2
-1	-4

(D)

x	y
0	-2
1	4
2	5
-1	-4

35. Which of these equations represents the table of values ?

x	-1	0	2	3
y	8	9	11	12

(A) $y = x - 9$

(B) $y = 9x$

(C) $y = 3x + 9$

(D) $y = x + 9$

36. The below box and whisker plots summerize the trial runs of rocket launch by team A and team B in minutes.

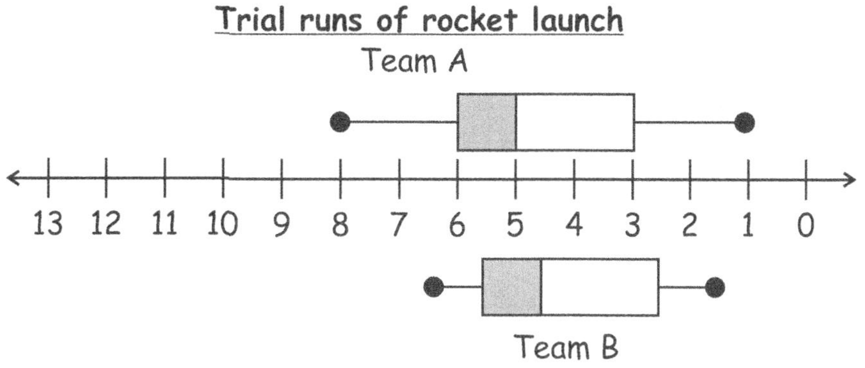

Based on the above data , which of the below statements is true ?

(A) The least time taken by Team B was 5 minutes
(B) The highest time taken by Team A was 10 minutes
(C) The least time taken by Team A was 1 minute
(D) The highest time taken by Team B was 7 minutes

37. The matrix below shows average class scores of grade 6 , grade 7 and grade 8 from 3 sections A , B and C in spanish midterm test. Which element represents the value of row 2 , column 3 ?

	Gr - 6	Gr - 7	Gr - 8
Sec - A	82	85	89
Sec - B	92	83	85
Sec - C	79	88	86

(A) 85
(B) 89
(C) 83
(D) 88

38. The below scatter plot describes the relation between the video game prices and the market demand (supply)

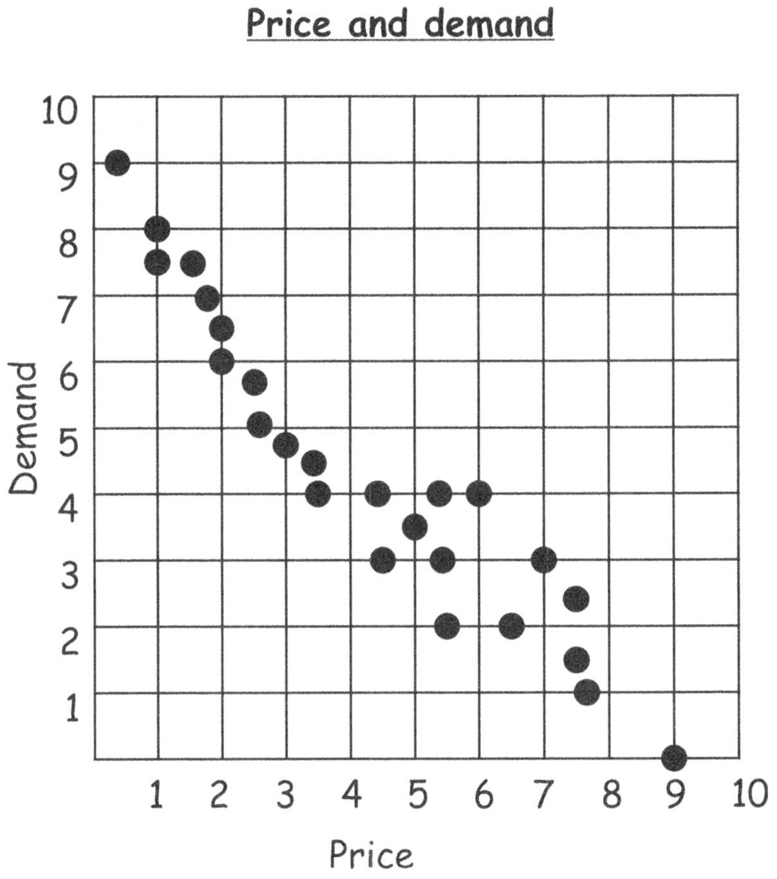

Based on the above data , which statement best describes the relationship between price and demand of the video games ?

(A) There is no correlation between price and demand
(B) Price and demand are equal
(C) Price increases , demand increases
(D) Price increases , demand decreases

39. A box contains 5 tiaras , 10 pendants , 8 charms , 2 bracelets. Jade picks an item from the box , what is the probability that it will be a tiara or a bracelet ?

(A) 28 %
(B) 25 %
(C) 31 %
(D) 7 %

40. Three friends share jokes with each other as given below.

- Mark shares 8 jokes with Greg and 13 jokes with Rob
- Greg shares 12 jokes with Rob and 9 jokes with Mark
- Rob shares 15 jokes with Mark and 11 jokes with Greg

Which of the below matrix summarizes the above information ?

(A)

	Mark	Greg	Rob
Mark	0	13	8
Greg	15	0	12
Rob	9	11	0

(B)

	Mark	Greg	Rob
Mark	0	8	13
Greg	9	0	12
Rob	15	11	0

(C)

	Mark	Greg	Rob
Mark	9	0	13
Greg	0	8	12
Rob	11	15	0

(D)

	Mark	Greg	Rob
Mark	0	11	12
Greg	9	0	13
Rob	15	8	0

41. Frank has $750 in his savings account. He pays $30 every week for a personal loan. Assuming Frank makes no withdrawals from the account. Which table represents his account statement ? (Excluding any interest amount associated)

(A)

Week	Account balance
0	$720
1	$690
2	$660
3	$640
4	$620

(B)

Week	Account balance
0	$750
1	$730
2	$700
3	$670
4	$640

(C)

Week	Account balance
0	$750
1	$720
2	$690
3	$650
4	$630

(D)

Week	Account balance
0	$750
1	$720
2	$690
3	$660
4	$630

42. This graph shows the number of bracelets sold at a trendy jewellary store in a week.

Day of Week	Number of bracelets sold
Monday	
Tuesday	
Wednesday	
Thursday	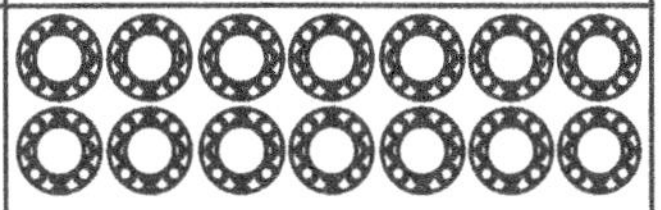
Friday	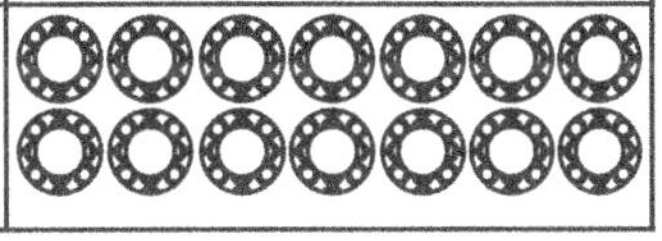

Each bracelet costs $4.00. What is the sale amount of the week ?

(A) $200

(B) $196

(C) $224

(D) $232

The bar graph given below shows the percentage distribution of the total expenditures of a company under various expense heads during 2019.

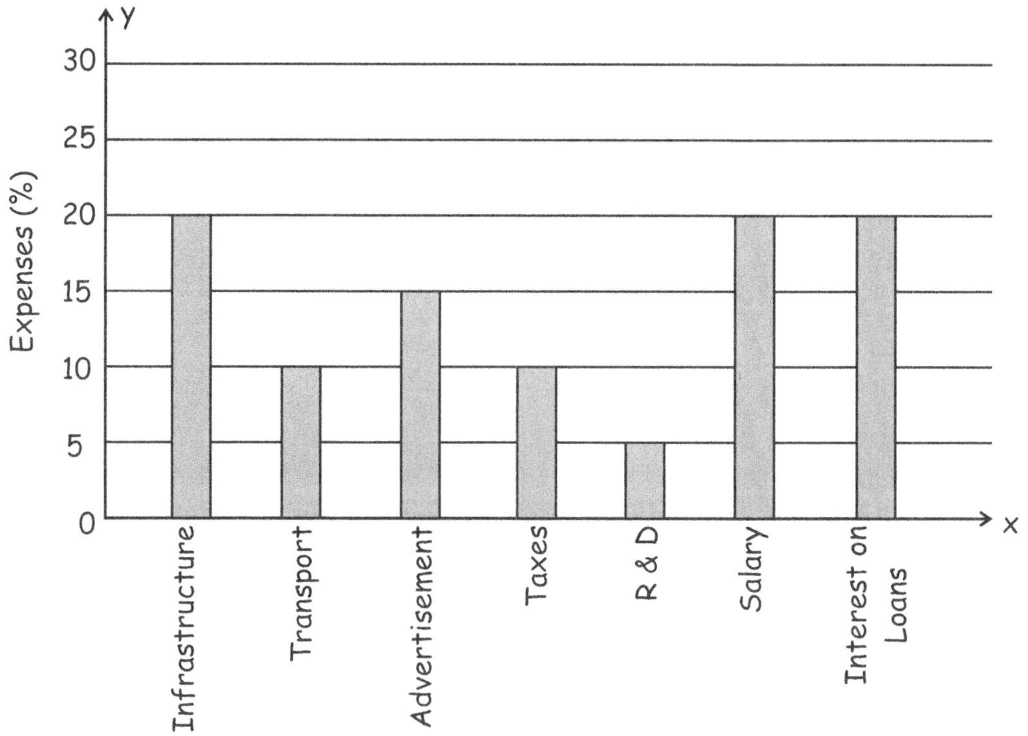

43. The total amount of expenditures of the company is how many times of expenditure on research and development ?

(A) 27 (B) 18

(C) 20 (D) 8

44. The expenditure on the interest on loans is by what percent more than the expenditure on transport ?

(A) 5% (B) 10%

(C) 20% (D) 50%

45. The expenditure on the infrastructure is by what percent more than the expenditure on advertisement ?

(A) 25% (B) 15%

(C) 30% (D) 22%

46. Evaluate $(4 \times 10^{-2})^4$ = ?

(A) 0.256×10^{-5} (B) 256×10^{5}

(C) 25.6×10^{-6} (D) 2.56×10^{-10}

47. Evaluate $\dfrac{860 \times 10^{-1}}{3.5 \times 10^{1}}$ = ?

(A) 30.1×10^{3} (B) 30.1×10^{-3}

(C) 2.457×10^{0} (D) 3.01×10^{3}

48. Evaluate $(4.8 \times 10^{-6})(3 \times 10^{6})$ = ?

(A) 1.44×10^{3} (B) 1.44×10^{1}

(C) 1.44×10^{2} (D) 14.4×10^{3}

49. The radius of a circular path is 10 feet. A 3 feet wide path is laid inside along its circumference. Find the approximate area of the circular path.

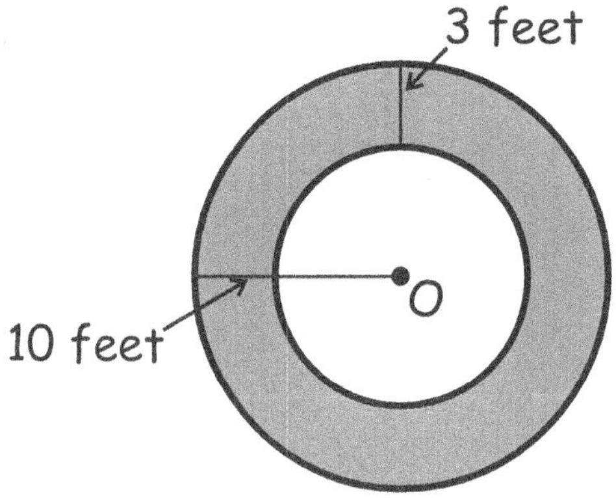

(A) 165 ft^2 (B) 160 ft^2

(C) 174 ft^2 (D) 138 ft^2

50. A wire in the form of a rectangle is of length 26 cm long and 11.68 cm wide. The wire is reshaped to form a circle. What is the radius of the circle (Note : ∏ = 3.14)

(A) 10 cm (B) 17 cm

(C) 9 cm (D) 12 cm

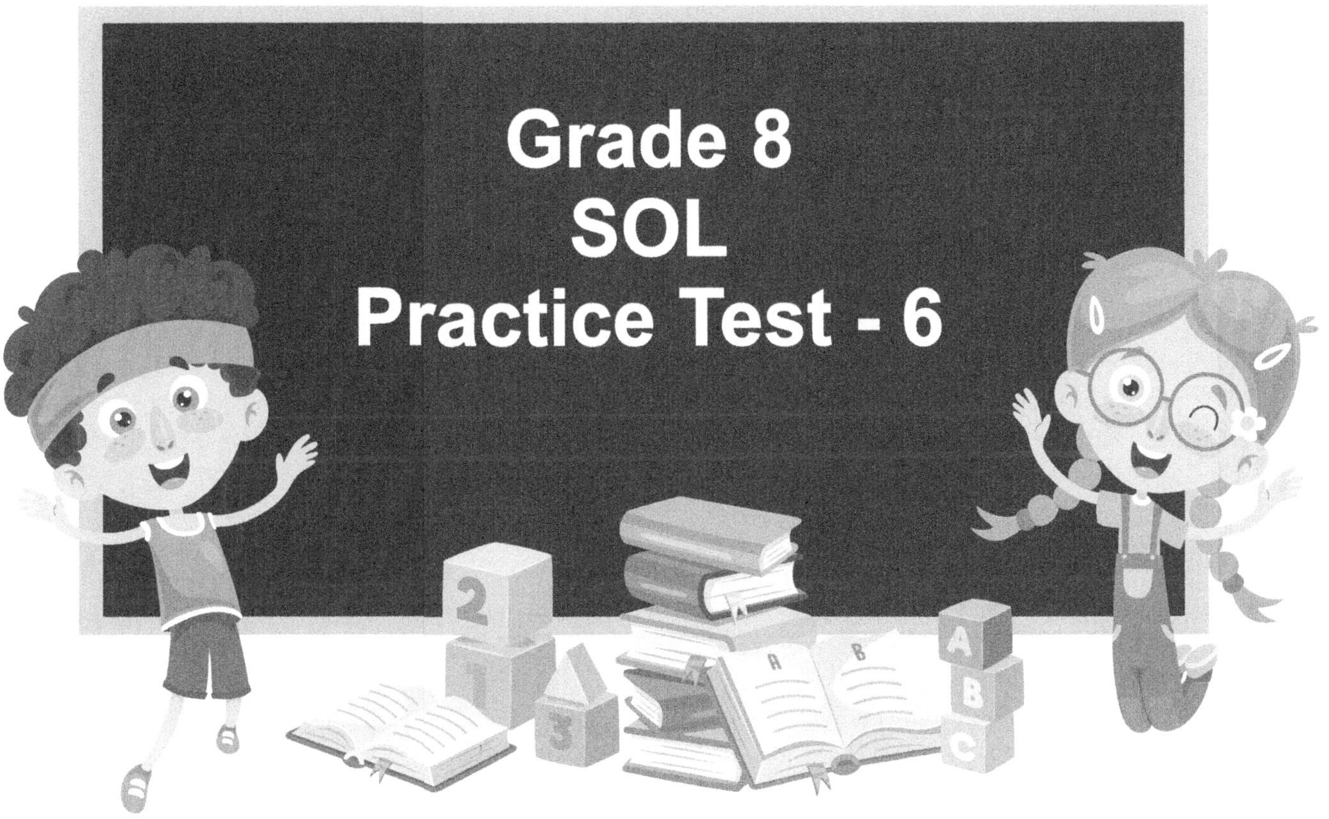
Grade 8
SOL
Practice Test - 6

1. Evaluate $19 - (9 - 48 \div (-16))$

(A) -10 (B) 1

(C) 7 (D) 26

2. Simplify $((-2)^2)^2 - (-4)$

(A) 20 (B) 14

(C) 33 (D) 5

3. Which of the below is less than 10101010 ?

(A) 10.1001×10^3 (B) 10.0101×10^7

(C) 11.11×10^6 (D) 11.01×10^8

4. Which of the below is an integer number ?

(A) $\sqrt{120}$ (B) -170

(C) $\frac{13}{19}$ (D) $\sqrt{120}$

5. Which of the below represents 0.07 in scientific notation ?

(A) 0.07×10^{-3} (B) 0.7×10^{-2}

(C) 0.7×10^{-3} (D) 7×10^{-2}

6. The set of natural numbers is not a subset of ________

(A) Irrational numbers (B) Integers

(C) Real numbers (D) Whole numbers

7. Which of the following describes a square root of 90 ?

(A) Between 9 and 10 (B) Between 10 and 11

(C) Between 45 and 46 (D) Between 89 and 92

8. Simplify $x^2 - ((-4) + y)$; where $x = 9$ and $y = 3$

(A) 75 (B) 84

(C) 82 (D) 80

9. Simplify $\frac{n}{2} - (m^2 + m)$; where $m = 9$ and $n = 2$

(A) -77 (B) -104

(C) -89 (D) -79

10. Which of the below is a perfect square ?

(A) 145 (B) 195

(C) 124 (D) 225

11. The square root of which of the following integers is between 7 and 8 ?

(A) 49 (B) 55

(C) 65 (D) 81

12. Violet purchased a cell phone X at $799.99 and sells it a 13% price increase. Find the selling price of the cell phone X.

(A) $140 (B) $959.99

(C) $919.99 (D) $903.99

13. Find the volume of a cone with a diameter of 12 yd and a height of 14 yd. Round the answer to the nearest tenth.

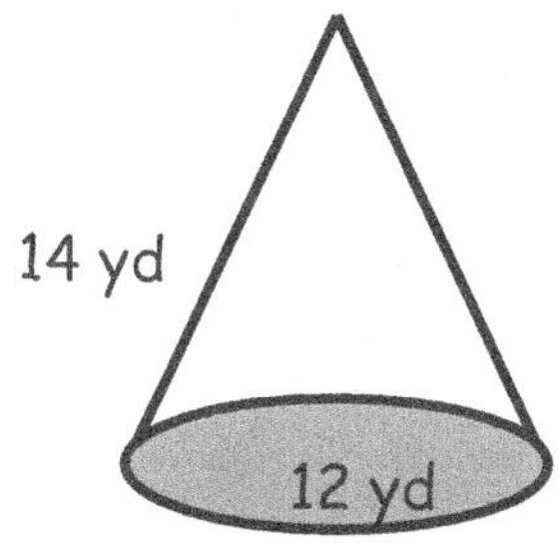

(A) 808 yd^3 (B) 453.9 yd^3

(C) 527.8 yd^3 (D) 404 yd^3

14. What is the relation between $\angle a$ and $\angle b$ from the below figure ?

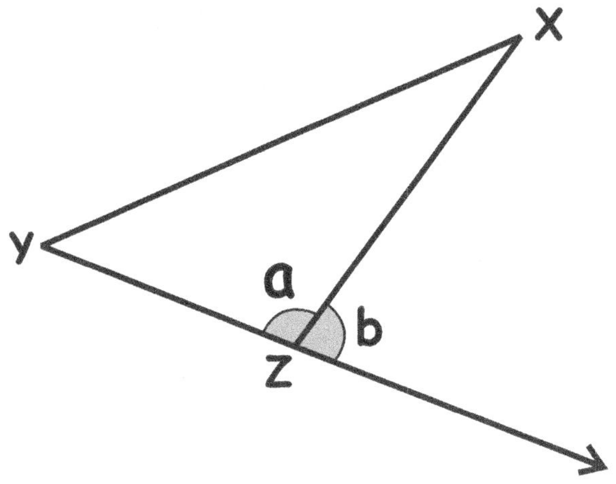

(A) Linear pair

(B) Complementary angles

(C) Corresponding angles

(D) alternate interior angles

15. Find the missing angle in the below figure.

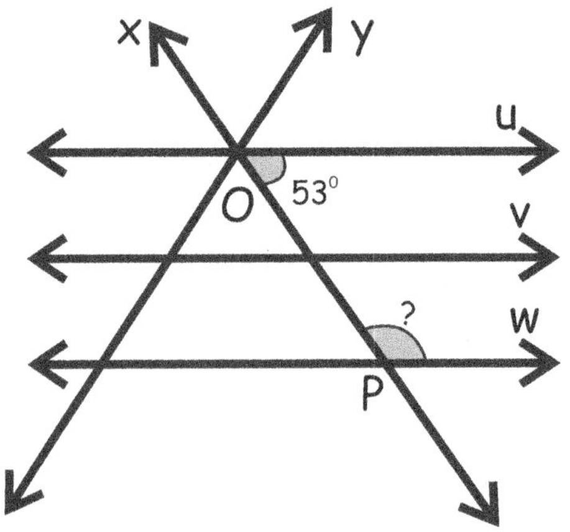

(A) 127°

(B) 115°

(C) 130°

(D) 117°

16. Which of the below options form a right angle triangle.

(A) 39 , 80 , 85   (B) 36 , 77 , 81

(C) 24 , 70 , 73   (D) 24 , 70 , 74

17. Which of the below options describes the transformation of the below figure ABC ?

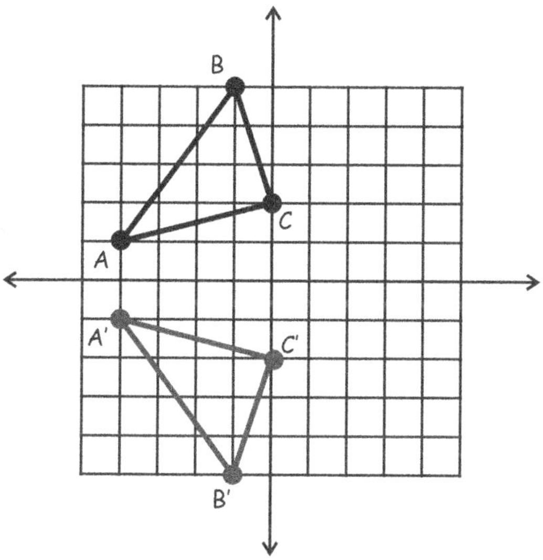

(A) Reflection across y axis

(B) Translation of 5 units left

(C) Reflection across x axis

(D) Translation of 3 units left and 2 units up

18. Find the volume of a square prism measuring 5 yd along each edge of the base and 4 yd tall. Answer to the nearest teth

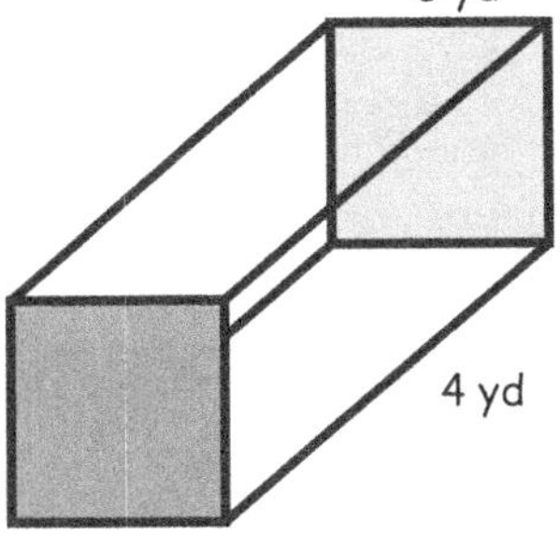

(A) 266.4 yd^3 (B) 200 yd^3

(C) 100 yd^3 (D) 222 yd^3

19. Which of the below is the hypotenuse ?

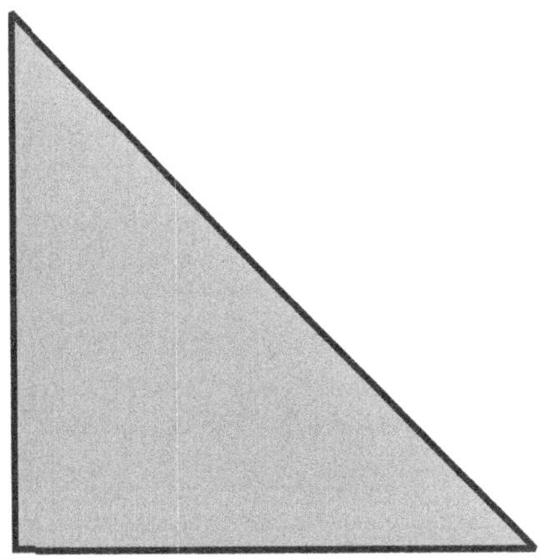

(A) $PQ^2 + QR^2 = PR^2$ (B) $PR^2 + QR^2 = PQ^2$

(C) $PQ^2 + PR^2 = RQ^2$ (D) $PR^2 + QR^2 + Pr^2 = 1$

20. Jack has a fair 6 sided dice. Which statement best describes Jade's chance of getting a 17 and win the game ?

(A) Certainly will win (B) Certainly will lose

(C) Most likely will win (D) Most likely will lose

21. Which of the below graphs represent the equation

$$y = x + 5$$

(A)

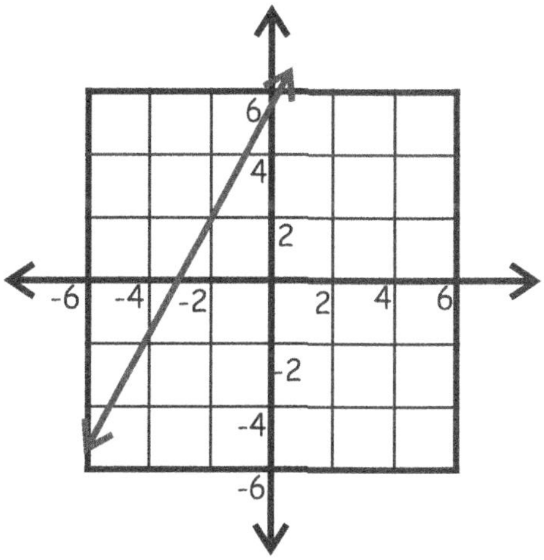

(B)

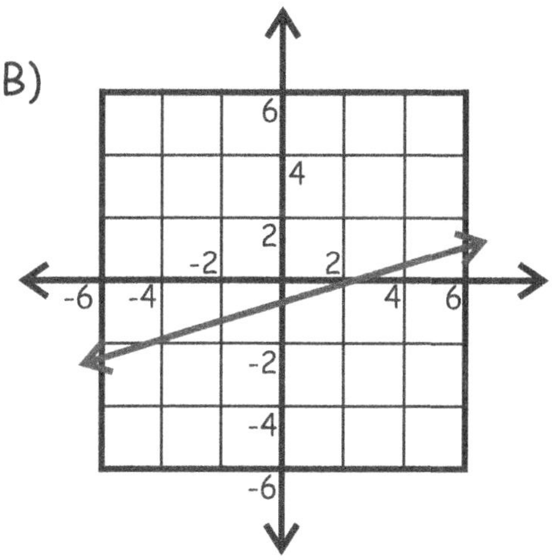

(C)

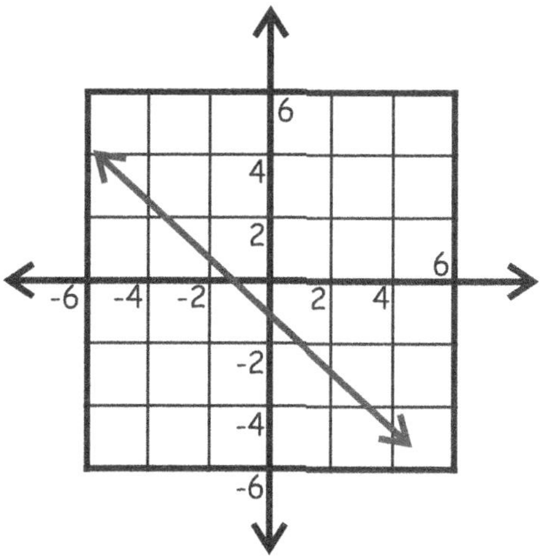

(D)

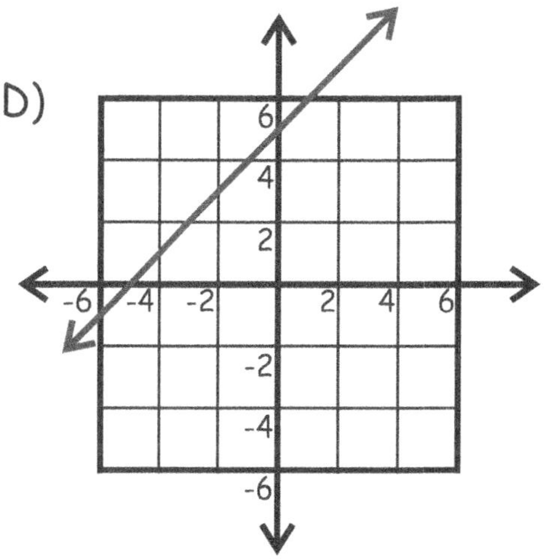

22. Simplify $-9x > -153$

(A) $x < -162$ (B) $x < 17$

(C) $x > 17$ (C) $x > -162$

23. Find the value of x $122 = x + 57$

(A) -65 (B) 179

(C) 65 (D) -179

24. Tina shared 6 candies with her friends. If she shares $\frac{3}{5}$ of all the candies she had , how many candies are left ?

(A) 3.6 (B) 10

(C) 11 (D) 4

25. The formula for the surface area of a right rectangular prism is $A = 2lw + 2hw + 2lh$, where l, w, and h represent the length, width, and height, respectively. Which term of this formula is *not* dependent on the height ?

(A) A (B) 2lw

(C) 2hw (D) 2lh

26. It takes Lisa 14 hours to clean an attic. Mia can clean the same attic in 16 hours. If they worked together how long would it take them ?

(A) 6.84 hours (B) 7.47 hours

(C) 5.79 hours (D) 8.25 hours

27. Simplify $-60 < -12x$

(A) $x > -48$ (B) $x < -48$

(C) $x < 5$ (D) $x > 5$

28. What is the range of the below data set ?

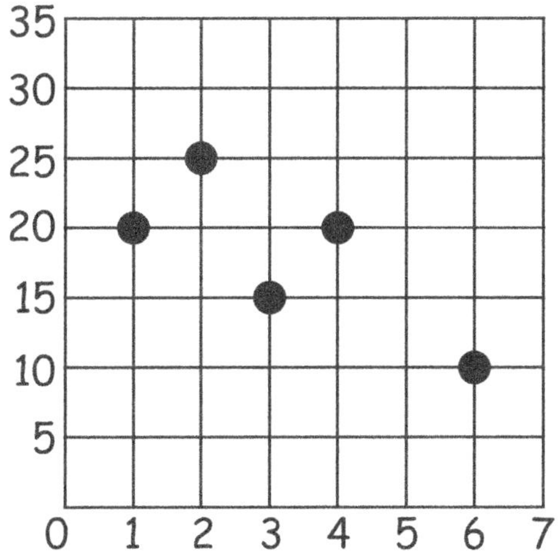

(A) {5,15,20,20,25} (B) {5,10,20,25,35}

(C) {10,25,30,35,40} (D) {5,15,25,35}

29. Kate has a savings of $1,000 at the end of January,2019.She pays $100 every month for her car loan. Which graph correctly displays Kate's savings ?

(A)

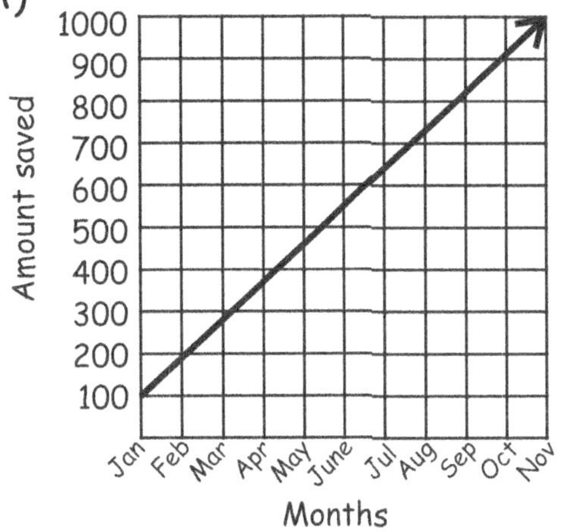

(B)

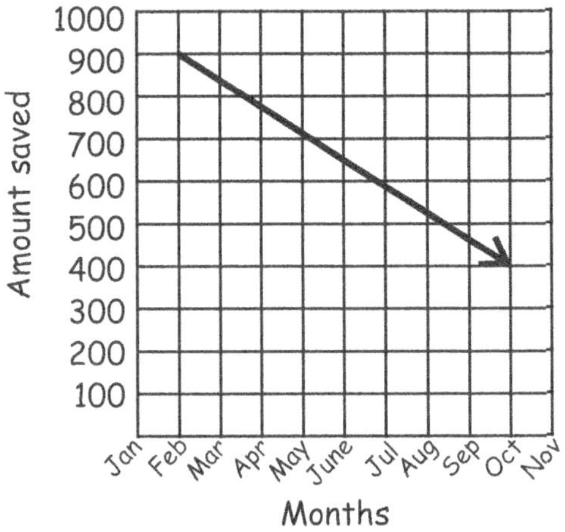

(C)

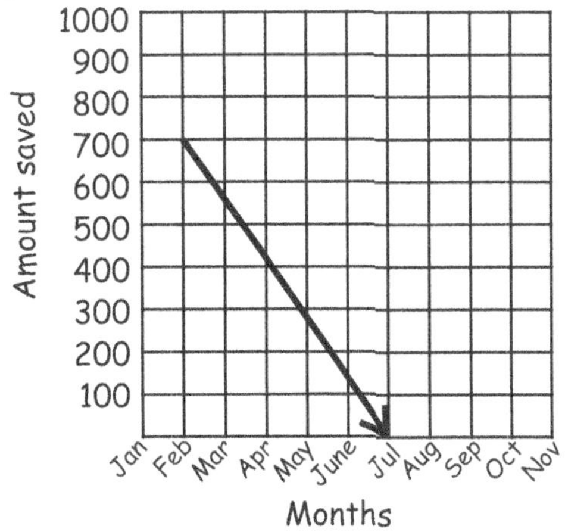

(D)

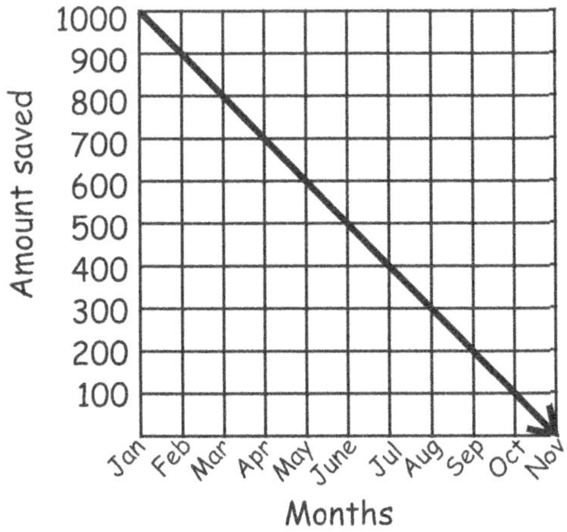

30. Which of the following tables of values is true for the equation

$$y = 6x + 1\ ?$$

(A)

x	y
-2	-11
-1	5
0	3
1	-7

(B)

x	y
2	-11
-1	-5
0	2
1	7

(C)

x	y
-2	-11
-1	-4
0	1
1	6

(D)

x	y
-2	-11
-1	-5
0	1
1	7

31. Which number is not an element of the range of the data given below ?

$$\{ (-7,9) , (9,10) , (-6,-5) , (4,-2) \}$$

(A) -5

(B) -10

(C) 9

(D) -2

32. Which graph contains the ordered pairs listed in the table ?

x	1	2	-1	-2
y	5	10	-5	-10

(A)

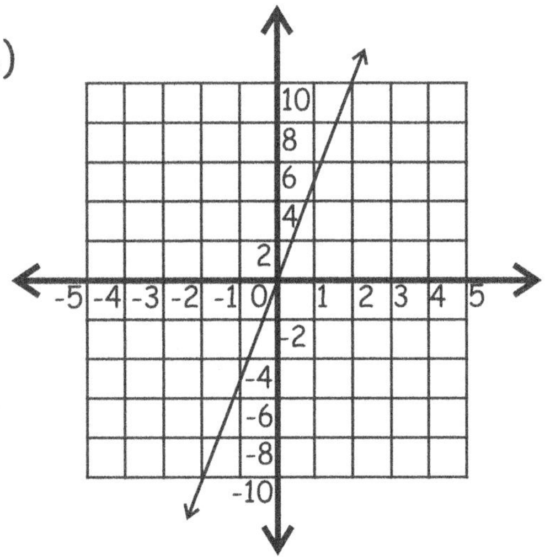

(B)

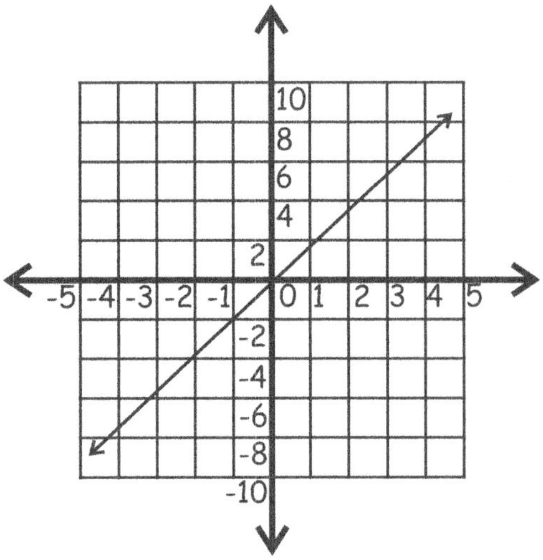

(C)

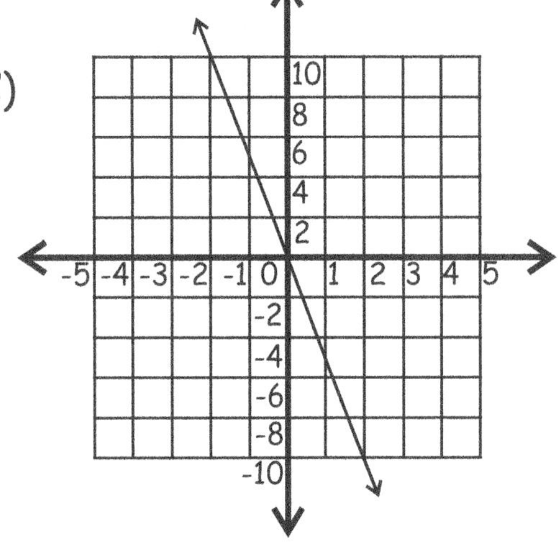

(D)

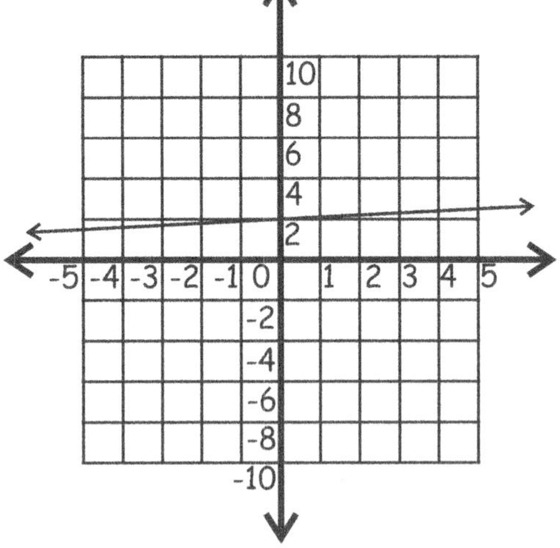

33. If the line containing the points in the table is plotted on a coordinate system , what does the graph look like ?

x	y
-4	-1
-8	-2
4	1
8	2

(A)

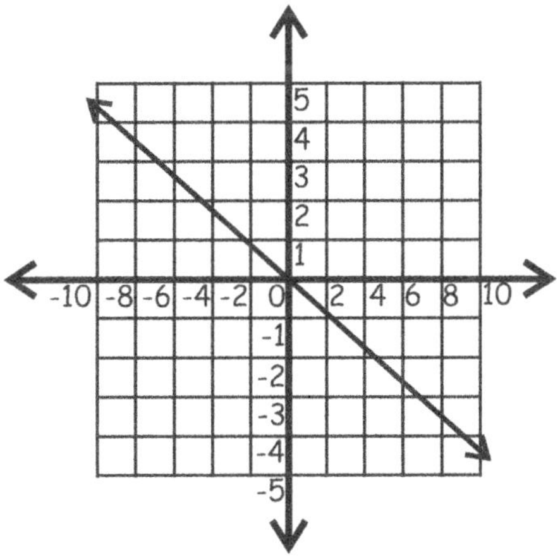

(B)

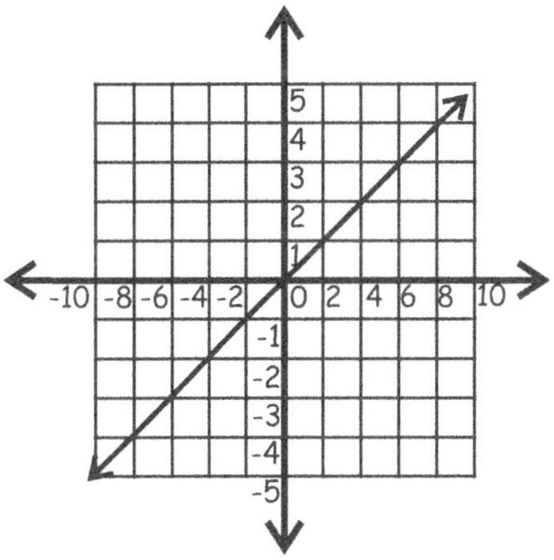

(C)

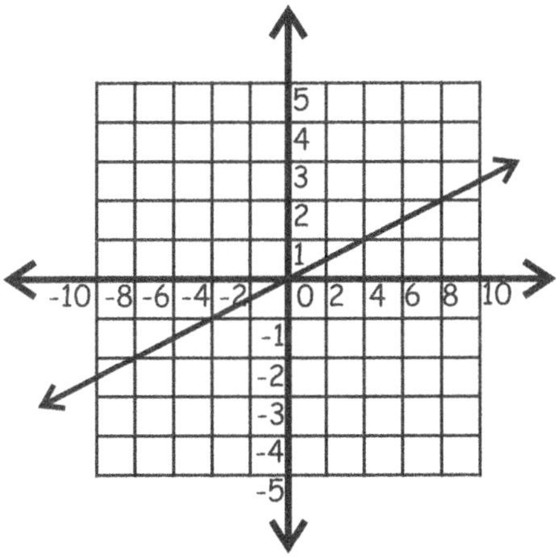

(D)

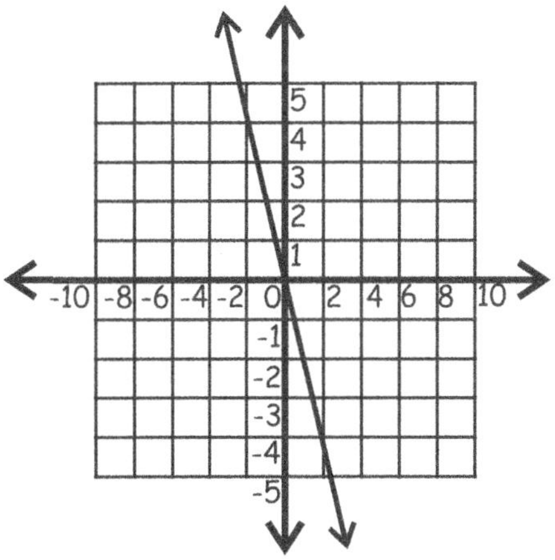

34. Which table contains only points tht lie on the line of the equation $y = 2x + 5$?

(A)

x	y
2	-1
2	9
4	11
0	-5

(B)

x	y
-2	1
2	9
3	11
0	5

(C)

x	y
0	0
2	9
3	11
6	5

(D)

x	y
-2	1
2	7
3	9
0	-5

35. Which of these equations represents the table of values ?

x	-1	0	1	2
y	-1	1	3	5

(A) $y = x$

(B) $y = x + 2$

(C) $y = 2x - 1$

(D) $y = 2x + 1$

36. The below box and whisker plots summerize geography bee scores of grade 8 and grade 7

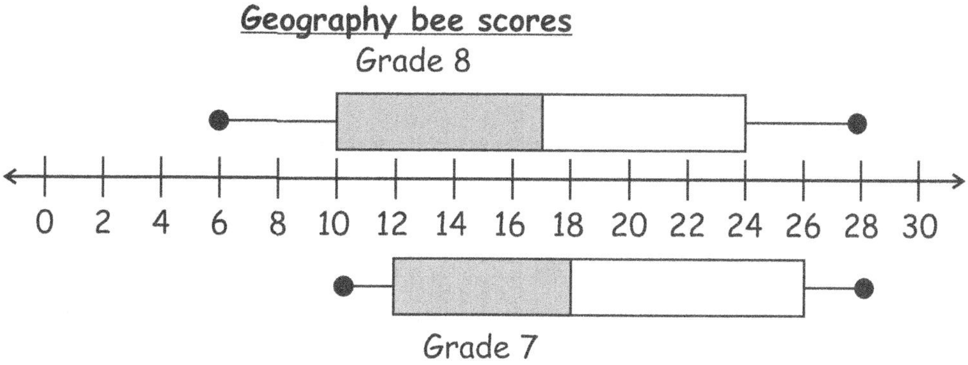

Based on the above data , which of the below statements is true ?

(A) The highest score of grade 7 and grade 8 students is 28
(B) The least score of grade 7 student is 6
(C) The least score of grade 8 students is 10
(D) No grade 7 students got a score of 26

37. The matrix below shows average class scores of grade 6 , grade 7 and grade 8 from 3 sections A , B and C in end of the year assessments. Which element represents the value of row 3 , column 1 ?

	Gr - 6	Gr - 7	Gr - 8
Sec - A	93	99	95
Sec - B	98	90	96
Sec - C	89	91	98

(A) 91
(B) 98
(C) 89
(D) 90

38. The below scatter plot describes the sales of burgers across the days at Yummy food corner.

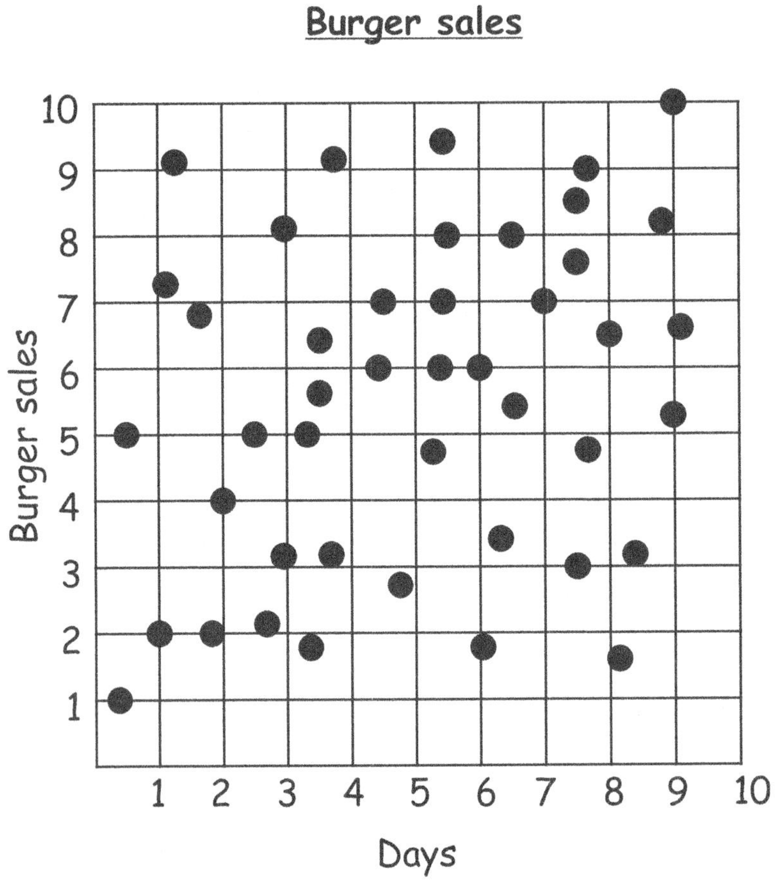

Based on the above data , which statement best describes the relationship between burger sales across various days ?

(A) Number of burger sales increased over the days
(B) There is no correlation between number of burger sales and days
(C) Number of burger sales decreased over the days
(D) Number of burger sales were same on each day

39. A box has 7 lego boxes , 4 toy robots , 9 drawing boards , 5 dart boards Noah picks an item from the box , what is the probability that it will be a drawing board ?

(A) 36 % (B) 40 %

(C) 28 % (D) 20 %

40. Three friends exchanged books with each other as given below.

- Taylor shares 4 books with Joey and 7 books with Linda
- Joey shares 9 books with Linda and 1 books with Taylor
- Linda shares 3 books with Taylor and 8 books with Joey

Which of the below matrix summarizes the above information ?

(A)

	Taylor	Joey	Linda
Taylor	1	4	7
Joey	1	0	9
Linda	5	8	2

(B)

	Taylor	Joey	Linda
Taylor	0	4	7
Joey	3	0	8
Linda	1	9	0

(C)

	Taylor	Joey	Linda
Taylor	0	7	4
Joey	1	8	9
Linda	3	0	0

(D)

	Taylor	Joey	Linda
Taylor	0	4	7
Joey	1	0	9
Linda	3	8	0

41. Dan has $1100 in his savings account. His mom deposits $100 every week. Assuming Dan makes no withdrawals from the account.
Which table represents his account statement ? (Excluding any interest amount associated)

(A)

Week	Account balance
0	$1100
1	$1200
2	$1300
3	$1400
4	$1500

(B)

Week	Account balance
0	$1200
1	$1250
2	$1300
3	$1350
4	$1400

(C)

Week	Account balance
0	$1100
1	$1200
2	$1400
3	$1500
4	$1600

(D)

Week	Account balance
0	$1100
1	$1200
2	$1250
3	$1400
4	$1500

42. The graph shows the number of evergreen trees sold each day during one week at a local nursery.

Day of Week	Number of ever green trees sold
Monday	
Tuesday	
Wednesday	
Thursday	
Friday	

Each tree costs $10.00. What is the sale amount of the week ?

(A) $200 (B) $250

(C) $220 (D) $210

The bar graph given below shows the percentage of car sales across various years. Based on the given data answer the questions 43-45.

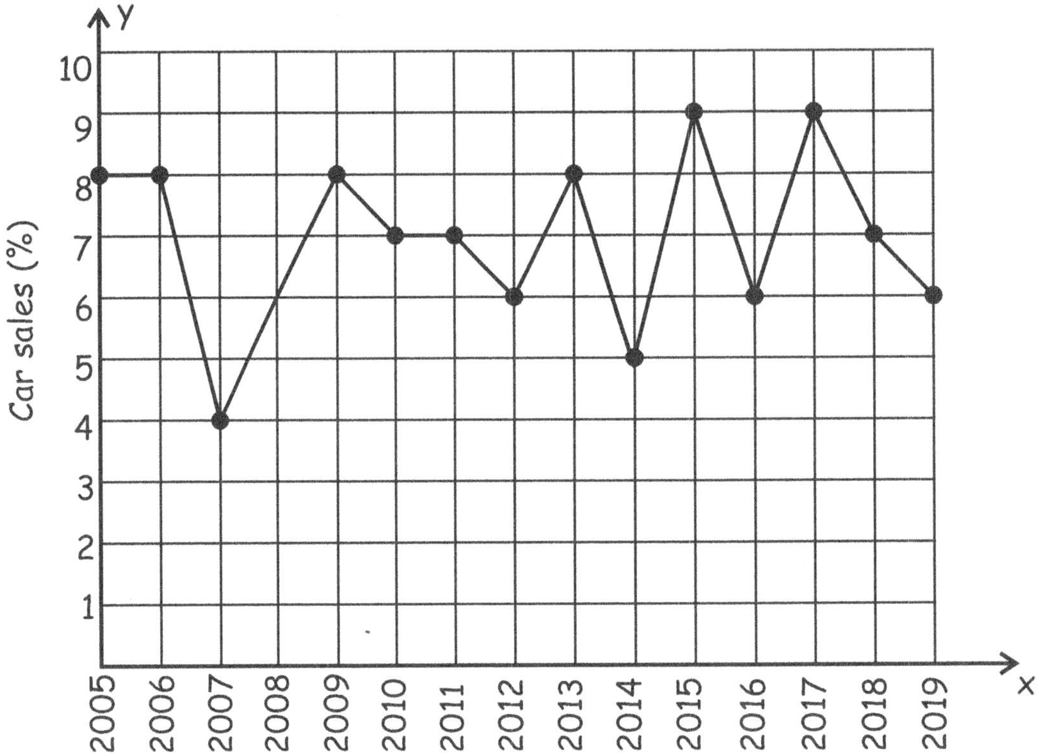

43. Find the mode of the data given above.

(A) 5 (B) 1

(C) 11 (D) 8

44. Find the median of the data given above.

(A) 12 (B) 7

(C) 13 (D) 5

45. Find the mean of the data given above.

(A) 5.72 (B) 7.09

(C) 7.00 (D) 6.59

46. Evaluate $(5 \times 10^{-5})^4 = ?$

(A) 6.25×10^{-18} (B) 62.5×10^{-18}

(C) 5×10^{-18} (D) 0.5×10^{-18}

47. Evaluate $\dfrac{2.85 \times 10^{0}}{4 \times 10^{-1}} = ?$

(A) 1.14×10^{0} (B) 7.125×10^{0}

(C) 7.125×10^{-1} (D) 1.14×10^{-1}

48. Evaluate $(9.65 \times 10^{-1})(7.3 \times 10^{1}) = ?$

(A) 7.045×10^{1} (B) 7.045×10^{2}

(C) 0.7045×10^{2} (D) 1.322×10^{-2}

49. The radius of a circular path is 18 feet. A 3 feet wide path is laid inside along its circumference. Find the approximate area of the circular path.

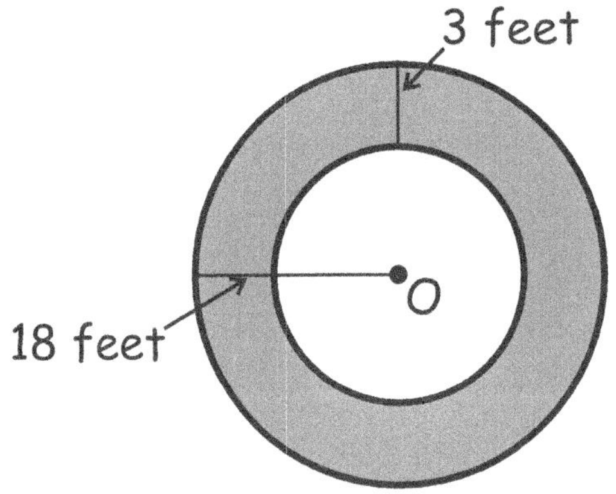

(A) 311 ft^2 (B) 309 ft^2

(C) 315 ft^2 (D) 319 ft^2

50. A wire in the form of a rectangle is of length 22.9 cm long and 21.06 cm wide. The wire is reshaped to form a circle. What is the diameter of the circle (Note : Π = 3.14)

(A) 25 cm (B) 32 cm

(C) 28 cm (D) 21 cm

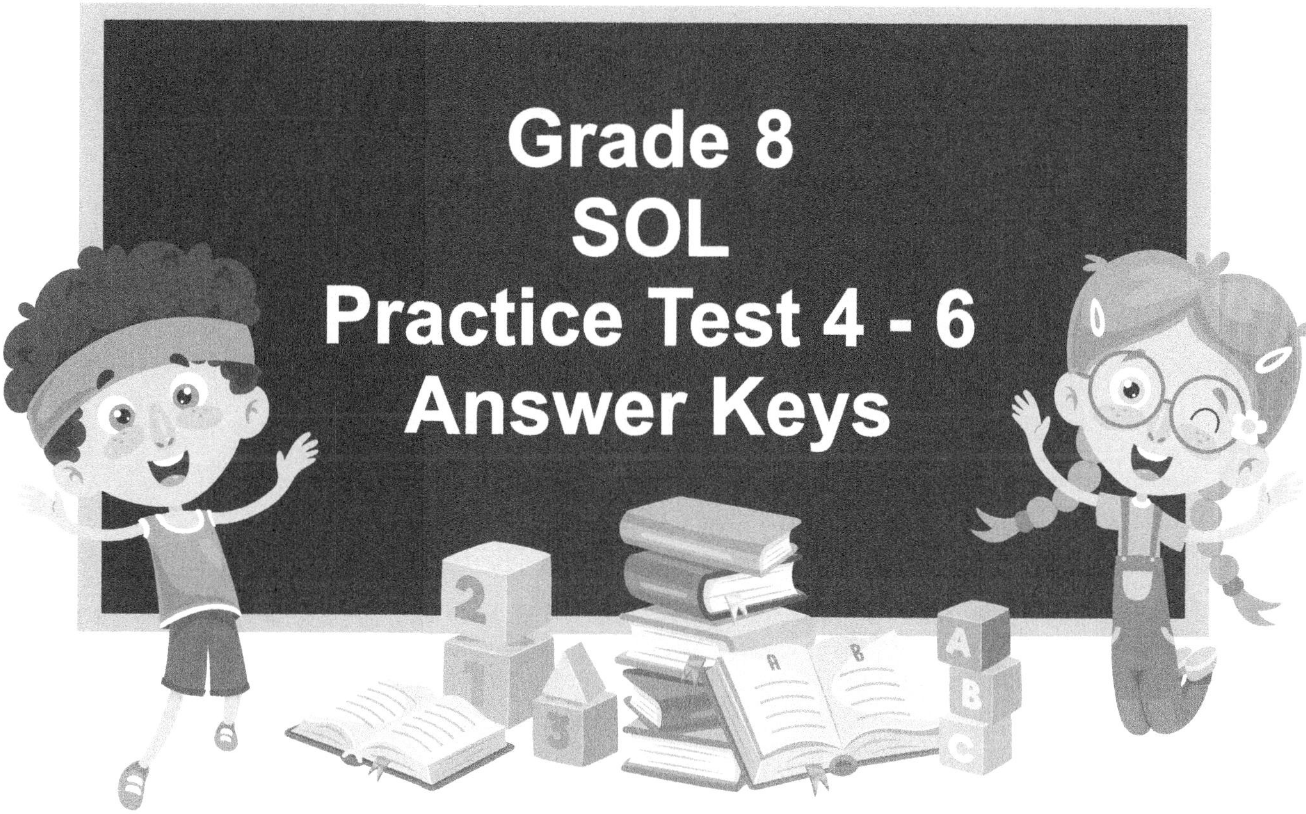
Grade 8
SOL
Practice Test 4 - 6
Answer Keys

Answer Key Test - 4

1. D
2. C
3. D
4. A
5. C
6. D
7. B
8. C
9. A
10. D
11. B
12. A
13. A
14. C
15. A

16. B

17. D

18. C

19. D

20. D

21. A

22. C

23. A

24. C

25. A

26. D

27. B

28. C

29. A

30. B

31. D

32. C

33. B

34. C

35. A

36. D

37. C

38. C

39. B

40. C

41. B

42. D

43. D

44. A

45. B

46. D

47. C

48. B

49. A

50. D

Answer Key Test - 5

1. A
2. B
3. A
4. D
5. A
6. D
7. C
8. A
9. B
10. C
11. D
12. A
13. C
14. A
15. D

16. C

17. A

18. D

19. C

20. A

21. B

22. B

23. A

24. D

25. A

26. D

27. A

28. B

29. D

30. C

31. A

32. D

33. D

34. A

35. D

36. C

37. A

38. D

39. A

40. B

41. D

42. A

43. C

44. D

45. A

46. A

47. C

48. B

49. B

50. D

Answer Key Test - 6

1. C
2. A
3. A
4. B
5. D
6. A
7. A
8. C
9. C
10. D
11. B
12. D
13. C
14. A
15. A

16. D

17. C

18. C

19. A

20. B

21. D

22. B

23. C

24. D

25. B

26. B

27. C

28. A

29. D

30. D

31. B

32. A

33. C

34. B

35. D

36. A

37. C

38. B

39. A

40. D

41. A

42. D

43. D

44. B

45. C

46. A

47. B

48. A

49. A

50. C

GRADE 8
SOL
Score Calculation

Score calculation

If you get this many times correct :	Then your converted scale scor is :
0	000
1	217
2	249
3	268
4	282
5	293
6	303
7	311
8	319
9	325
10	331
11	337
12	343
13	348
14	353
15	357
16	362
17	366
18	371
19	375

SCORE

20	379
21	383
22	387
23	391
24	395
25	399
26	403
27	407
28	411
29	415
30	419
31	423
32	427
33	432
34	436
35	441
36	445
37	450
38	456
39	461

40	467
41	473
42	480
43	487
44	496
45	505
46	516
47	531
48	550
49	582
50	600

Made in the USA
Las Vegas, NV
30 September 2026

55324682R00063